Wallace, James

A voyage to India

Inktank publishing

Wallace, James

A voyage to India

Inktank publishing, 2018

www.inktank-publishing.com

ISBN/EAN: 9783747768754

A

VOYAGE TO INDIA:

CONTAINING

REFLECTIONS ON A VOYAGE TO MADRAS AND BENGAL, IN 1821, IN THE SHIP LONACH;

INSTRUCTIONS

FOR THE PRESERVATION OF HEALTH IN INDIAN CLIMATES;

AND

HINTS TO SURGEONS AND OWNERS OF PRIVATE TRADING-SHIPS.

BY JAMES WALLACE,
SURGEON OF THE LONACH.

The dullest writer, if he holds up virtue and liberty, and benevolence and esteem; if he fills his story with good and wholesome advice; and dresses up his little magazine of knowledge or entertainment, at least with a good intention; he may make himself useful.

CITIZEN OF THE WORLD.

LONDON:
PRINTED FOR T. AND G. UNDERWOOD, 32, FLEET STREET;
AND CHALMERS AND COLLINS, GLASGOW.

MDCCCXXIV.

DEDICATION.

IN MEMORY OF

THE LATE JOHN PEARSON, Esq.

Commander of the Lonach;

AND TO THE

OFFICERS, PASSENGERS, AND CREW

OF THAT SHIP.

A CONSIDERABLE time before these "Reflections" were given to the press, he under whose auspices they were written, and who would have patronised them, had he been living, was numbered with the dead. I can now, therefore, only dedicate them to his memory; and well indeed does it deserve the tribute at my hand. This is not the place to speak of his goodness—of the ties that were between us; to tell how much he did for me—and how much I owe him; yet I may be allowed to state, that it is for the kindnesses he heaped upon me, from the first moment I came under his command, till the hour when I saw his eyes closed in death, that I now sorrowingly bring forward his name. Nor is it I alone that regrets him ; all who sailed with him—all who knew him—will, I am confident, when they remember Captain John Pearson, remember him with affection and respect, and shed a warm tear at his early fate.

To my other shipmates in the Lonach I also dedicate the volume. They are now scattered far and wide ; but it probably may meet the eye of some, and possibly may awaken in their bosoms some fond recollections. It, at all events, carries to them my affection and regard unabated.

JAMES WALLACE.

CONTENTS.

PART I.

PART II.

PART III.

PREFACE.

It may be a matter of surprise to some, the appearance of "reflections" on a voyage so common as that to Madras and Bengal. We have of late, however, had "Travels," and "Voyages," and "Tours," given to the world in abundance, speaking of parts and things perhaps not more interesting or new than an Indian voyage. Nor do I think, for all this travelling and publishing, that the ground is yet wholly occupied, or even the most common of subjects altogether exhausted; there seems still to be a little room left, though so many have been already forward to tell their stories, for another straggler yet to step into and tell his. But indeed I pretend not to give much new in what I now present; the reflections, I

b

believe, being just such as must naturally occur to the majority of those who embark on a sea voyage. None, however, that I know of, had before thrown them into the form which I have here adopted. When I entered on my first voyage, like many others, I meant to have kept a private journal of daily occurrences, merely for my own satisfaction and amusement. But it soon occurred to me, that a volume such as the present, might be acceptable to a few out of the many who annually take their departure from England for India. It occurred to me, that a thing which would give the traveller to India some idea of what he is to meet with on his journey to it, as well as on his arrival; a thing which might instruct as well as amuse, might be useful. I therefore relinquished my first intention, and set about the present work. And though in writing it, I kept India and my Indian voyage principally in view, I wrote what I thought might also apply to any voyage. My object has been, to give an idea of some of the most striking circumstances to be met with at sea; to paint the feelings of the traveller journeying to a foreign land; and I hope though the Indian voyager will find what has been said especially to apply to him, even the voyager to any country will find something in it worthy of his perusal.

The first part of the work, "The Reflections," I have divided into four chapters. In the first chapter, "The Departure," I have attempted to describe the varied feelings experienced at our first entering upon a voyage; the thoughts called up at bidding adieu to our native soil; with some of the miseries of an embarkation, *&c.* The second, "The Voyage," contains a detail of some of the most obvious and interesting occurrences which are presented to the notice of the traveller on the deep; a history of some of the circumstances best calculated to instruct and amuse in the profession of the mariner. In the third, "The Approach to the Indian Shore," besides other

b 2

allusions, I have touched chiefly on the state of mind we are likely to possess as we draw to the foreign coast. I have attempted to shew what are, and, perhaps still more, what ought to be, the feelings at such a period. The fourth, and last, "The Arrival," is taken up with what is experienced actually on reaching the foreign shore; with some few remarks on India and its inhabitants, allusions to home, *&c.*

And I hope what I have touched upon, I have portrayed tolerably correctly. It will be seen that I have attempted to vary the matter a little; that I have not attached myself alone to the serious, but have endeavoured to mix up a little of the gay with it. And in all parts, I have made it my object to impress morality and proper conduct; wherever I could, I have brought before the reader the lesson of religion, and pointed the mind to the Great First Cause. The poetry intermixed with the volume, taken from some of the most esteemed authors, and adapted as appropriately as I could to the subjects, will also add something to its interest.

In the second part of the work, "Observations for the Preservation of Health in Indian climates," I have given such directions, I think, as will be quite sufficient for any one taking up his residence in these climates. They are plain and few; but they are the most important; and he who has them impressed properly on his mind, I should hope will find them in some degree advantageous to him. They apply, as the reflections do, chiefly to India; but of course they must apply also to all climates like it; to the west as well as to the east; the advice given in them I trust may be worthy a remembrance almost in any climate; for in no situation is health to be preserved without care.

At the commencement of this division, I have thrown in, by the way, a few observations relating to consumptive patients. They may call the attention of some to the remedy advised, who otherwise might neglect it.

The third and last part, " Hints to Surgeons and Owners of Private-trading ships," I should also hope will have its use. It will give the young man entering upon that employment some considerable idea of what he is to meet with, and prepare him for it; it will give him a few hints which perhaps may be of some little service to him. The remarks, too, which apply to the owners of these ships, may not be altogether lost. It will be to the advantage of themselves, as well as to the advantage of every one else, that what I have spoken upon be attended to and remedied. And I am satisfied it needs only to be stated, to have the remedies applied.

In short, as it is I give it to the world. If it has any thing to recommend it, the public will find it out, and give it its due. If it be undeserving of attention, and if I have judged wrong in ever letting it see the light, it will of course be allowed to sink into the mass of idle tales that have gone before it, and be forgotten. Nor will I be a great sufferer by the sentence; for as a writer I have no reputation to lose; a mere adventurer as yet upon the sea of authorship, should the wind prove at once unfavourable, I can easily put "about ship," and get back to my old port again.

A VOYAGE TO INDIA,

&c. &c.

PART I.

CHAPTER I.

THE DEPARTURE.

—— The pilgrim on his journey bent,—
Oft will he pause, and gaze the journey'd plain;
Oft pause again, the valley to survey,
Where food or slumber sooth'd his wand'ring way.

SOUTHEY.

On Friday, the 9th of March, 1821, the preparations for sea being finished, and crew and passengers on board, the Lonach sailed from Gravesend for Madras and Bengal; and on Sunday, the 11th, anchored in the Downs.

When we enter upon a new scene, and get introduced amongst a people altogether strange to us, we naturally feel a little gloomy, even though it be a scene of our own choosing, and the people we get amongst, in every respect,

B

as suitable as strangers can be. It takes a time ere the mind can accommodate itself to any change it may experience; even allowing the change to be from worse to better. It takes a time after our introduction to strangers, however affable they may be—although, indeed, they may have dispositions exactly suited to our own—before we can find out what the disposition really is, or feel ourselves altogether at home in their presence. And thus it is, that for the few first days, on board ship, the young adventurer is by no means so happy as he could wish to be. He stalks about in the midst of confusion; unable, though he may be willing, to apply himself in earnest to any thing. There are many others around him, but he scarcely knows them yet; the formality and distance of a first introduction have not yet given place to the cheering familiarity of ripened friendship; and, even in the midst of them, he feels lonely. And thus left, as it were, to himself, his thoughts naturally turn to that which interests them most. He looks back on happier days; he ponders on the friends he is leaving; he glances forward to the future—but he sees not there, to cheer and comfort, what he finds in the past; doubts and fears fill his bosom; and if in his disposition there exist much sensibility and tenderness, he is indeed apt to become melancholy.

But with every evil there is an accompanying good; there is no situation of distress without its accompanying circumstances of alleviation. The great object to be attained, when the mind is tending to a desponding subject, is properly to occupy and divide the thoughts: and, fortunately, the very confusion we get amongst when we go first on board ship, though it interests us but little—though, indeed, we may think it rather tending to our annoyance—serves as a kind of stimulus to the drooping mind; it arrests and feeds, though we may not be aware of it, the then depraved thoughts, as we may well term them—thus preventing them from settling too steadily on any one subject, and drawing us insensibly away from the desponding revery.

Very soon, too, we meet with some one, who, feeling as we do ourselves, is inclined to shake off melancholy, by a participation of thoughts and cheerful conversation; and this, at any rate for a time, enlivens not a little the feelings. And, as we every day get rather more acquainted with the people, we every day get a little more reconciled to the place; the spirits gradually rise, and we soon begin to think our situation may not be so intolerable as we at first imagined.

But it is not at the very first that the tender feelings are most powerfully aroused. Although he who embarks for a far country, may be melancholy enough at the very time of embarkation, yet, if he supposes that at this time his feelings are at the worst, he mistakes; he has still severer pangs to experience. The tie which binds him to his home, though it may be stretched, is not yet broken. A little time must elapse ere this takes place; and it is just at the breaking of it, that his feelings of sorrow are wound up to the highest.

On the morning of Tuesday, the 13th, the wind, which, from the period of our reaching the Downs, had continued unfavourable, came round; and early the following morning we weighed, and with fine weather stood down channel;—and it was then that I experienced my saddest moments. It is when we feel ourselves fast leaving the happy land of our nativity, and fairly out of the reach of all its enjoyments, that the mind, in spite of every effort we can use, feels the touch of melancholy. It is then that the value of a home is found out—that we ascertain what a blessing it indeed is, to have kindred and companions to associate with—that we get fully convinced of the truth of the saying, "it is only when our comforts are gone we learn the value of them." By a law of nature our affections are entwined around the place of our birth. Where we have been born and brought up, where we tasted our earliest, sweetest joys, and where those who are dearest to us remain, there our affections cling, and from thence there is no tearing ourselves without suffering many a pang. Cold indeed must the heart of that man be, who, as the vessel, borne along by the steady breeze, is carrying him farther and farther from his native shore, can refuse to it his constant lingering gaze, or stand to see it recede without emotion. At such a time the man of feeling is melted into grief. When he calls to mind the days of playful youth—the parents and brothers and sisters, who had so long fondly loved, and anxiously tended and watched over him—the companions he associated with—the happy scenes he mixed in,—and feels, that now for a long time, perhaps for ever, he is leaving all—his bosom becomes the seat of many a rending passion, and the sigh that is struggling in his breast, and the tear that is bursting from his eye, he finds it impossible to restrain. Ay, and even he who never thought he possessed such kindly and patriotic feelings—he who, up to the hour of trial, may imagine he has a heart quite stout enough, to give up all its pleasures, and break with all its associations, without regret, will find, when the hour comes, he has calculated wrong; and this man, with all his hardihood, will be seen, like others, as the vessel is fast speeding along, and the high summit of the land has almost sunk into the wave, eagerly endeavouring to catch another glimpse of that which probably by him shall be seen no more.

Come hither, hither, my little page,
  Why dost thou weep and wail?
Or dost thou dread the billow's rage,
  Or tremble at the gale?

Let wind be shrill, let waves roll high,
  I fear not wave nor wind;
Yet marvel not, Sir Childe, that I
  Am sorrowful in mind:
For I have from my father gone,
  A mother whom I love,
And have no friend, save—

Enough, enough, my little lad,
  Such tears become thine eye—
If I thy guiltless bosom had,
  Mine own would not be dry.

Yet, though it is pleasing and amiable to indulge in these feelings to a certain degree, it is not right to carry them too far. There is an apathy—a cold-heartedness—that would teach us to stand unmoved under every circumstance; and there is an over-great sensibility, which would unman, and make us sink at the most trivial afflictions; neither of which is it necessary to acquire. But there is a medium state, which it is possible to find, and which enables him who finds it to stand firm; while at the same time, he sympathizes, to feel, in their proper force, his own or his neighbour's sorrows—while, at the same time, he is capable, if necessary, of withdrawing his mind from the subject of sorrow, and fitting it for the performance of other duties.

And, if we examine, we will find, how natural, and how happy it is, for the wanderer to gain such a state of mind. No doubt, when he feels himself taking leave of all his comforts; parting with those that are dearest to him on earth—with the nearest and the kindest of his kindred—with the friends and companions of his earliest days; in a word, bidding a long, perhaps an eternal, farewell, to all that ever interested his affections,—his heart must be filled with many an anxious and distressing thought. He possesses something more than firmness—something not so amiable as firmness—if he does not find his spirit bend, and sink in some degree, at such a separation. If there be much affection and kindness in his nature, he is indeed apt to bend too much to the afflicting load, and shew that he wants even that degree of firmness which it is right he should have. But when, again, he remembers, that after a time has elapsed, he may return to his happy home and the bosom of his friends,—when he thinks that absence will only ripen and increase the affection, and that when again he meets those he is now parting with, and tastes those pleasures he is now forsaking, his happiness will be tenfold; when he feels assured, that wherever he goes, the prayers and affections of many go with him, and that, should it be his lot to return, his return will be greeted by many a fond and anxious heart—there is much consolation given to the mind—much of its anxiety is removed. By proper reflections, the feelings are gradually balanced, and the mind comes to gain just the strength that is wanted. Hope administers to the wanderer under all his troubles. Her taper, burning to the last hour of life's pilgrimage, mixes up with the darkest of his feelings a gleam, which ever keeps despair from the borne-down spirit—which cheers and supports him in all his wanderings.

These, however, are only the troubles of the mind; and though they are quite enough, yet, as they generally wear off, or at least lose much of their poignancy, as soon as the parting is fairly past, and even while they last are pleasing though painful, if there were no others, there would not be much cause for complaint. But there are also troubles of the body, as well as the mind, to be endured; and these, unfortunately, are all pain, without the least intermixture of pleasure. Just as we are getting over the sorrow of parting—just as our spirits are on the move to their wonted level, and we are beginning to promise ourselves something like enjoyment and cheerfulness again, there comes upon us a sickness, a most deadly sickness, which, at once, sinks the spirits still lower than before, and reduces the body still lower than the spirits, and most cruelly puts an end to every hope and prospect of comfort.

During the whole of Wednesday, the day we left the Downs, the wind continued fair and moderate, and there was no appearance of complaint amongst us. On Thursday, the weather was still fine, and for that day also, I may say we were all well enough. Towards evening, indeed, some felt a little lightness of head, and rather a dimunition of appetite, but they did not think it a matter of much importance, and imagined a sleep would put all to rights. In the course of the night, however, the breeze increased considerably, and the swell of the sea, and the motion of the ship along with it, and those who in the evening had their spirits tolerably well up, had them miserably enough depressed in the morning; those who in the evening were promising themselves an early start, found, when the morning came, they had neither inclination nor ability to get upon their legs. I felt, I believe, about as uncomfortable as any of them, but I resolved to make a bold push, and get up to see what was going on above. So out of bed I got with some difficulty, and after managing, with more, to get on my clothes, I made my way to the deck, just as breakfast was ready. At the call to breakfast, I went in, although I had little inclination for the meal, to take my seat at table along with the rest; but I found few others there; few, out of our number, had made their appearance, and even these few, as soon as the eating articles were produced, and the caterer had commenced sending them round, got up and made their exit rather hurriedly. I followed very soon, for I found things far from being right, and, like the others, took the way to my cot, into which I tumbled in a most pitiable condition.

All Friday it continued to blow, and all that day, I and many others, were, as we supposed, in a state of greater misery than ever mortal was in before. The moment we raised our head from the pillow, such horrible sensations ensued, that we were glad instantly to get it down again. Some who fought against their feelings, and got out of bed, had scarcely reached the floor, when a tumble of the vessel sent them tumbling to the other side of their cabin; and before they had time to recover themselves another roll would send them as quickly and roughly back again; and thus tumbled and bumped about, they were glad to get into their softer and safer birth again. Others who got the length of beginning to dress, in the attempt to draw on a stocking, or in any other act which occupied both hands, and put the body on rather a ticklish balance, were thrown down with such violence, that for some days afterwards they had cause to remember it. And to increase the misfortune, the same lurch that upset the man, generally upset along with him, some of the cabin furniture, and not only did he run the risk of sustaining an injury from the fall, but it was a wonder, indeed, if he did not also get a salute from a trunk, or a dressing-box, or a basin, or something or other rather too ponderous in its nature. And some, after managing to get themselves wholly equipped, became all at once so intolerably ill, that they were absolutely obliged instantly to tumble into bed again, clothes and all. And what added to our distress, in all our troubles, there was no one pitied us. When the spirit is sunk, from whatever cause, the voice of sympathy can do much to restore it. When the hand of disease is upon us, how enfeebled is its grasp, if the soothing attentions—the cheering consolations of kind friends, are aiding us! Yet, although we imagined our situation just one calculated to excite the greatest commiseration, it drew forth no pity—it seemed not to be viewed thus by any one else. From the tried sailor, a laugh or a swear was all the comfort we could get. And even those who were as young in the life as ourselves, but who luckily had escaped the terrible malady, seemed also to be without any feeling of sympathy; for though they came occasionally to visit us in our distress, they neither shewed that anxiety, nor gave that condolence, which we, judging from our feelings, thought we deserved. They generally started off in a very few minutes with the excuse, that if they stayed longer they would get sick too.

Burns says, in his " Address to the tooth-ach"

> When fevers burn, or agues freeze us,
> Rheumatics gnaw, or colics squeeze us,
> Our neebours sympatheese, to ease us
> Wi' pitying moan.
> But with thou—the hell o' a' diseases,
> They mock our groan.

And admirably, I think, is this adapted also to sea-sickness.

Neither did Saturday bring us any relief; indeed, we were fully worse that day than the one preceding. Towards the afternoon, when we thought, with sorrowful hearts, we had left the English shores for a long time behind, the wind, which, till then, had kept fair, came right against us, and increased to a gale. The swell of the sea, too, became great; the rain descended in torrents ; the bitter unfriendly wind kept howling through the rigging; and every now and then a big wave struck, and broke over the vessel. The whole scene, indeed, was a most dismal and disheartening one,—just calculated to sink still farther, our already too-far depressed spirits.

I shall always remember the night well. As I lay in the evening, in my cot, wishing myself any where but where I was, quite worn out both in body and mind, and careless what became either of myself or the ship, if I was only allowed to remain at rest, a message came that one of the men was taken ill, and that it was necessary I should see him. It was a most disagreeable piece of intelligence. Much rather would I have gone twenty miles at any other time to visit him on shore, than get up that night to go twenty yards. But there was no getting off well. So up I got: and having reached the deck, I proceeded cautiously along, groping my way in the midst of the storm, in quest of my patient. But just as I had nearly reached the place, a tremendous sea came gush in where part of the bulwark had been carried away, and half drowned me. Says I to myself—if this be the kind of practice I am to have now, it matters not how soon I am done with it. However, that was no place for reflection; I found out the sick man, and found the way back to my own birth again. But still the work was not done; other misfortunes still awaited me. My patient required a little medicine; and after I had, with no small difficulty, weighed out the different ingredients, and was just on the eve of putting past my weighing apparatus, and parcelling up my medicine, the vessel took a roll, and my cot, which was hanging up, taking one with it, struck me so forcibly that I was thrown from my seat, and away went the scales one way, and the weights another, and the paper with the medicine went off with the rest, and all my labour was to do over again. In sorrow, I ejaculated, as I lay on the floor amid the general ruin,—if a man could be in search of such a thing as misery, let him go to sea and he 'll be sure to find it: who is it, that either for curiosity, or gain, or glory, would lead such a horrible life? who is it that would not prefer the scantiest pittance on shore to the greatest affluence here? Let me be once more upon dry land, and if ever I set my foot on board ship again may I never get out of it. But complaints were vain; so after a few such melancholy reflections, I gathered up my scattered materials, and, setting to work again with rather more caution, got the business finished.

All Saturday night we fought against the storm, but finding it impossible to make head against it, on Sunday morning we put about ship, and stood back for the land. My sickness still continued as bad as ever, and had now reduced me almost to skin and bone; but though I was scarcely able to move, when I heard the glad tidings, that we were again making for the land, and likely to get relief, at least for a time, to our troubles, I once more made my way to the deck; and there I witnessed a most melancholy scene. I was not, I soon found, the only one that had suffered; for there I saw the greater number of those I had left on Thursday night, in good spirits, with ruddy countenances and masculine figures, most miserably changed. The fine clothes, before so clean and neatly fitted, were now only half put on, and largely coated over with blanket-wool.

The fashionable-tied neckcloth, and the spruce done up hair, were no longer to be seen. The bloom of the cheek was altogether gone; the energy of the eye had fled with it; the features were shrunken; and this, with the long black beard, which had evidently been allowed to grow in peace for some days, gave to the whole countenance, nay, to the whole man, a most woful and haggard appearance. The spectacle altogether was, indeed, a lamentable one; as we thus stood, upon the deck in silent sadness, eagerly bending our eyes in the direction of the land, the place where, alone, we looked for comfort.

Well might a spouter have said, as Macbeth,

> What are these,
> So wither'd, and so wild in their attire,
> That look not like the inhabitants o' the earth,
> And yet are on it! Live you, or are you aught
> That man may question?—This is a sorry sight.

On Sunday evening at six o'clock, after many ineffectual attempts and hard struggles, we got in to Falmouth harbour; and so great a dislike to a sea life had I impressed upon me at that time, that really I had considerable doubts, even situated as I was, whether I should ever leave terra firma again. I verily believed that I was always to be the victim of sea-sickness; and, consequently, that at sea I never could have any comfort. Nay, I made sure, that had we remained at sea but a few days longer, I should have been sent out of the land of the living altogether. And thus impressed, I certainly, on Sunday forenoon, as we were bearing in upon the coast, had, as nearly as possible, made up my mind to stay upon it when I got to it, and forego India for ever, if a ship was to be my conveyance to it.

But it is the nature of man to forget his griefs, and not always to stand by his resolves. The very sorrows which, at the moment of his suffering them, he imagines so great as certainly to leave a most lasting impression, in the course of no very great time, are hardly remembered; and the resolutions which now he lays down most absolutely, he very soon sees it good to alter. And it is good for him that he is so formed. And so it was with me at Falmouth. A day or two on shore recruited me wonderfully; and the bitter remembrance of my previous unhappy situation, faded considerably. Every one, too, was crying out, that we had now got all our sickness over, and that however long we might be at sea, we would have no more of it. And as both honour and necessity urged me to go through with what I had bound myself to perform, I listened to the cry, and forgetting all my former resolutions, resolved at last, whatever might be the consequences, to try myself once more.

And accordingly on Wednesday the 21st, when the ship sailed again, I sailed with her; and again I was launched into a sea of affliction. Soon were all my fine hopes of immunity for ever from the unwelcome complaint overturned; for the morning after we sailed, although I got out of bed in tolerable spirits, determined, if possible, to brave every thing, I was very soon glad to give up the contest, and turn in again. And for a whole week after, I continued, with many others, in a state, which I sincerely pray, no poor soul may ever be in. At last, however, there came a gale,—a gale which had almost driven us in on the Spanish shore, and, probably, finished our sickness not exactly in the way we wanted; and whether it was that that frightened us into a recovery, or good weather which we had immediately after, that brought about the change, I will not determine positively ; but, at all events, from that day there were evident symptoms of amendment, and very soon all complaint disappeared.

Regarding a cure for sea-sickness, I am afraid there have none been found out yet. Many plans have been recommended both for its prevention and cure, but none, I believe, have ever been found of much service. Some tell us to remain always on deck, exposed to the air. But I can tell them, from experience, that it is a thing hardly possible to keep up the head long, when once we are fairly sick ; and more, that it is a thing very far from agreeable to remain a whole day on the deck of a ship, in the English channel, in March or months like it. Others advise us never to get up at all,—to keep our bed constantly till we feel ourselves well. But this is just running into another extreme, and is not advisable. A third set recommend what would be rather an agreeable remedy, if we had any relish for food, namely, to eat, continually. Eat, say a great many, heartily and often, whether you have an appetite or not, and you will find it a most effectual cure. But this I am inclined to regard as the very worst of advices. It is quite impossible that the stomach, weakened and worn out as it always is in the complaint, can digest food in any quantity. It may be made to receive it, but it must of necessity reject it again; and such a plan will assuredly tend more to aggravate, than to lessen the complaint. Unless there be some inclination for it, solid food should not be taken. Others go again to the opposite extreme, and bid us let eating alone altogether, till we are well. Others recommend different kinds of spirits, aromatics, *&c.* But the truth is, the motion of the ship is the cause of the malady, and until we can find a way to prevent this, I am afraid those who are disposed to it, must be content to suffer the ill for a time. It is only when the system has accommodated itself to the cause, that it ceases to influence.

But although I think it hardly possible, by any means, either to prevent or cure, I think a little may be done to alleviate. And the advice I would give for this purpose is,—to sit up, and in the open air, as much as possible,—and when absolutely obliged to lie down, immediately on feeling a little recruited, make another attempt at sitting up. Eat but little solid food, even though somewhat inclined for it; and when there is no inclination, or perhaps, rather a dislike to it, let it alone altogether. It is, however, very generally the case, that there is some particular kind of food which the stomach would relish, though it loaths every other; and, of course, if this can be had, it will be very suitable. Although every kind of solid nutriment may be loathed, there is, very frequently, a desire for spirits, or wine, or something of that sort, prepared in particular ways. The stomach—the whole system—debilitated and sinking, calls for something to arouse it a little,—to keep its vigour up, till it masters the complaint; and therefore, not only when the desire exists should it be gratified, but even when it does not exist, if the complaint is continuing for any length of time, some nutritive stimulus of this kind, should certainly be had recourse to. I felt myself, I know, from a little brandy and water, though brandy was a spirit I never could bear before, the very best effects. I think it right, however, to mention, that the stimulating system must not be too freely indulged in. Though I had always disliked brandy before, yet finding such considerable relief from the first dose, I repeated it, as I always do with any medicine I find answer the purpose, and continuing to take it till the end of my sickness, I found by the time I had fairly recovered, that the medicine was by no means disagreeable to me; and ever since I have been able to take a little of it. This is a hint which I hope will not be overlooked.

But patience, after all, is the best cure. A little time will, I may say invariably, remove the complaint; and what is some consolation, when we do get over it, we are frequently better than we were before.

CHAPTER II.

THE VOYAGE.

> Lo! to the wintry winds the pilot yields
> His bark, careering o'er unfathom'd fields.
> Poor child of danger, nursling of the storm,
> Sad are the woes that wreck thy manly form.
> But Hope can here her moon-light vigils keep,
> And sing to charm the spirit of the deep!
>
> CAMPBELL.

As yet I have only been shewing the darker side of a sea life, or voyage; and I dare say I have said almost enough to deter any one from going to sea, if by any means he can keep on shore. But I must also shew something of its bright one; and I think what I have to say upon that head will go a good way to counterbalance what has been said before.

And first, I would observe, that the description I have given of the troubles we suffered, from boisterous weather, at the outset of the voyage in the Lonach, will not apply to every voyage. It is only now and then that a ship, at starting, meets with such weather as we were exposed to; and, of course, it is only now and then that such sorrowful scenes are to be witnessed. Frequently is the ship, from the time she leaves her port till she has travelled far, till she has reached a more equable and genial clime, wafted along by the gentlest gales upon a quiet sea; and the system thus gradually initiated into the business, feels none of those disagreeable effects which so generally attend upon boisterous weather. And there are some who are not at all disposed to sea-sickness, who are fortunate enough to possess a constitution proof against it in all weathers.

And even when it does happen that, at the very outset, a storm is encountered, it lasts only for a time, and serves this good purpose, that when better weather does come, we value and enjoy it the more. Whatever may be the state of matters at the beginning, we may make ourselves sure, barring accidents, that after getting on a little, we will get into fine weather; and it is then that our dislike to the situation, however much it may have been, is sure to give way, and we begin to see in it much more comfort, many more sources of enjoyment, than the first tasting of it promised. By this time every one on board has become better acquainted, and frequent association and agreeable conversation, which are the soothers of life in all situations, become general. The mind is now, in some degree, weaned from the objects it has left, and begins to take some interest in those that are before it. The very business of the ship becomes attractive and amusing; and often is the mere passenger, who has nothing to do with it, in a very short time, almost as well versed in sea phrases, and able to talk with nearly as much freedom about braces and bunt-lines, and sheets and yards, as the seaman himself. Often is the young fellow, who, but a few days before, lay in a state of misery, wishing himself a thousand miles away from every thing like a ship, as ready as any one to lend his hand to a rope, and perhaps the very first to jump up to the yard, in a blowy night, when a top-sail is to be reefed or furled. But this last action, as it is dangerous, and is not necessary, is not to be commended.

I have heard it often said, that a life at sea is a most disagreeable and monotonous one. I have heard the complaints often made, that while we are on board ship, the time hangs exceedingly heavy; that there is no variety, no room to roam about; that day after day, and week after week, is the same; that we are just pent up in a prison, and to make our confinement worse, we are continually exposed to numerous dangers and privations and causes of anxiety. And I allow, that in many cases, the life is disagreeable enough; for in all things much depends on circumstances. I must allow the complaints to be true, to a certain extent, but I cannot admit them in their wide acceptation: for I know that not a few have just been as happy at sea, as ever they were on shore. I have felt that, even on the deep, far removed from the land, with all its varied enticing scenes and busied inhabitants, there is quite enough to keep away the complaint of tediousness; that there are objects well fitted for the contemplative mind; that with all the dangers and privations, and confinement, there are sources of enjoyment.

At sea, in a well-conducted ship, every thing is carried on with the greatest regularity. Certain times are set apart for certain things to be done; and he that is inclined to apportion his time and employ it properly, may certainly do it there without any difficulty. The cabin is an excellent place for retirement, and reflection, and study. And when there are a few others whom, when we please, we can converse and associate with, surely we cannot say, that the time can hang heavy. He that cannot find enough of occupation in his book and his friend, will scarcely find it, I am afraid, much to his profit, in any thing else. Yet it is not necessary that we should confine ourselves to the cabin and study alone. The mind certainly requires diversion, and it need not be without it at sea. When we feel inclined, we can do there just as we do on shore—go out and make our calls. We can pay to our friends, as regularly there as any where else, the morning or afternoon visit, and have the visit returned; and he only who has made one of it can tell the happiness which exists in the little society, when a few, with dispositions alike, are met in the little cabin, either conning with each other the lesson of knowledge, or joining in some innocent amusing recreation, or talking over joys past or joys to come. There we are secure from the bustle and allurements of the world; there the mind is induced to seek in itself for something to satisfy; there we learn to prize the few objects we have, and to hold fast in affection the heart that beats responsive with our own. Before dinner, or in the evening, we can take, as on shore, although we cannot range just so far, our promenade. We may close the evening, as we are inclined, either with music, or cards, or dancing. And when Saturday night comes, and all are assembled, with bounding hearts we drink to sweethearts and wives, and the cheerful song passes round. In truth, it is just a little city—a little society—of itself; and secluded as it may seem, he who likes to accomodate himself to it, need not want comfort.

One complaint I have said is, that privations and dangers attend it; and I admit that they do. The seaman, in particular, who has charge—who has duty to perform—has indeed enough of anxiety and privation. But there is no situation without its dangers; and though we may allow that the man who is tossing about on the wide ocean, runs, upon the whole, a greater risk than him that is walking quietly on terra firma, yet, we need not look very narrowly to discover, that when out at sea, though we may be exposed to dangers enough, we are out of the way of not a few that on land we would be liable to. In the various circumstances and situations of this uncertain state, it is indeed hard to say who stands the best chance. Fortunately, he who has to endure some of its privations, appreciates best its sweets when they come.

D

Another complaint is, the sameness of the scene. There is nothing, it is said, pleasing or beautiful to attract the eye; probably for months together nothing is to be seen but sea and sky; and, consequently, the scene can have few charms—the prospect must be tedious and uninviting.

But these are assertions which I must also regret, in a great measure. That for months together nothing else is to be seen but sea and sky, is true; but to tell me that in this sea and sky there is nothing beautiful or attractive, is to tell me what experience has proven to be false. It is true, that the mighty precipice, towering in majesty—and the mountain, with its less gigantic, but milder and more alluring aspect—and the valley, with its murmuring stream and its peaceful peeping cottage—and the flowery meadow, with all its variegated richness—and the wide-spread lawn, covered with its green verdure—and the far-extending forest, robed and blooming in its green livery—objects which give such life and pleasure to the eye and heart of man—are wanting; but it does not follow that, though these are absent, the prospect should either be tedious or uninviting; these are not essential to an attractive scene. I give them all their due; I give them, in their place, my warmest admiration. But I must still maintain, that even on the face of the deep, far away from all the loveliness of the vegetable world, there is much to attract and much to admire; and that he who possesses a mind capable of admiring Nature in any of her forms, will find enough to interest it there. It is there that the great orb of the day is seen, in all his splendour, slowly to rise from the bed of the ocean, and after travelling his course through the cloudless heavens, in the evening, to sink again into the wave; and as he slowly measures his course, and his bright beams, dancing on the bosom of the playful ocean, have gilded the surface of the green rippling wave, and the whole space of æther is touched with his splendour and serenity, surely there is much of beauty in the scene. It is there, in the tropical region, that when the bright luminary has finished his course and just gone down, and his beams, still darted upward from the deep, have tinted, in all the variety of colour, the clouds that occasionally rest on the verge of the horizon, a scene is produced, lovelier by far than any that is to be found in the whole circle of nature. And though it lasts but for a little—though the beams are soon gone, and the scene dies—yet, when the Queen of Night has taken her place in the heavens, and the stars, spread out upon the wide concave, are all sparkling and burning around her—while the meteor is giving forth its occasional blaze—another not less pleasing is brought forward. And he that, in the quiet hour of the night, when nought is to be heard, save the gentle murmur of the breeze, and the grateful noise of the bubbling wave, as the vessel, steadily pursuing her course, turns it aside and cuts her way through the deep,—and, perhaps, the voice of the seaman upon watch, as, to chase the lingering hour, he paces the deck, and converses with his comrade, or chants at the cheering, patriotic, or love ditty,—

he, I say, that can contemplate at such an hour without delight—he that would give his imagination play upon such a scene, and yet say that his emotion did not rise with the contemplation of it: that he could not dwell upon it with increasing delight, even till it disappeared—I pronounce him not to be alive to the charms of nature; he knows not how to estimate her beauties at all.

Nor should we complain that the scene remains the same; for, occasionally, it is indeed wonderfully and grandly changed.

" There was a tempest brooding in the air,
Far in the west—above, the skies were fair,
And the sun seemed to go in glory down;
One small black cloud (one only), like a crown,
Touch'd his descending disk, and rested there——."

" The sky is chang'd!—and such a change! O night,
And storm, and darkness, ye are wondrous strong,
Yet lovely in your strength, as is the light
Of a dark eye in woman——."

Nature, as if wearied of her long-continued smiling aspect, puts it aside for a time, and appears in her wildest and most striking attire. The golden day and the spangled night disappear. Dark lowering clouds gather and roll, and sweep along the vault of heaven, and by and by cover the whole concave. The gentle breeze begins to gather strength, and the far-spread sea loses its gently-undulating surface; above, beneath, all around, Nature has hung her dark portending mantle. And now the clouds, which threatened, burst, and send down deluges of rain. And now another element comes in with its awful aid—flash follows, in quick succession, flash of the vivid lightning, and peal comes after peal of the rolling thunder. The mighty blast sweeps along, and every moment increases; while the convulsed sea, rolling with the most terrific fury, seems ready to whelm and bury for ever, in its yawning gulf, the struggling diminutive bark. The whole elements seem engaged in some terrible trial; there seems to be a mustering up of their whole force, a striving, as it were, who shall send forth the greatest strength, and gain the mastership;—and some dread event—destruction—ruin—seems at hand. And then, indeed, may he whose delight it is to contemplate the grand workings of Nature—the various and arresting operations of the Almighty—find a theme well fitted to occupy him. Then may he of the wild and solitary mind, whose affections are not with men, nor with the enjoyments of social life, but whose occupation is to look for ever with the eagle eye of musing and searching curiosity, and eagerly to grasp at whatever is *outré* or attractive in the wide range of creation,—give full scope to his moody imagination. Like the bard, whose delight it was "to retire from his companions, and bury himself in the recesses of his native woods, or to ascend some eminence during the agitations of nature; to stride along the summit while the lightnings flashed around him, and, midst the howling of the tempest, to apostrophize the spirit of the storm"—may we find enough to attract—enough upon which to exert all the energies of mind—in such an awful and striking scene.

"Roll on, thou deep and dark blue ocean, roll!
  Ten thousand fleets sweep over thee in vain;
Man marks the earth with ruin—his control
  Stops with the shore;—upon the watery plain

The wrecks are all thy deed—nor doth remain
A shadow of man's ravage, save his own;
When, for a moment, like a drop of rain,
He sinks into thy depths with bubbling groan,
Without a grave, unknell'd, uncoffin'd, and unknown.

" Thou glorious mirror, where the Almighty's form
Glasses itself in tempests!—in all time,
Calm or convuls'd, in breeze or gale or storm,
Icing the pole, or in the torrid clime
Dark heaving—boundless, endless, and sublime!
The image of Eternity!—the throne
Of the Invisible! Even from out thy slime
The monsters of the deep are made! Each zone
Obeys thee! Thou goest forth dread, fathomless, alone!"

But, a little while, and all is calm again. Soon are the winds subdued, and the dread convulsion finished. All that threatened is removed, and again there is the gentle steady gale, and the serene cloudless sky, and the little beaming curling wave.

And let us just go on to give an example or two of the lessons that are taught, of the instruction which we may draw, from what we have attempted to describe. Is not the picture at sea, just the picture of human life in general? The bright sun rising, and soon setting again, shews man how he also enters on the morning of life, and reminds him of the short course he has to run; and the tinted evanescent horizon, is a beautiful emblem of his transient fast-fading joys. Like the favourable gale, which steadily and quietly drives the ship on her course,—like the serene prospect all around, which gives such hope and peace to the mariner's breast,—and the tempest coming suddenly, and threatening to destroy,—are all the affairs of humanity. Man, though he may travel for a time in the road of prosperity, can never make sure of continuing in it through the whole of life's journey. However promising and settled-like his prospects, he is never secure against the inroads of adversity. And just at the moment when the pulse of hope beats highest,—just when he thinks himself most safe, and hardly fears a single danger,—the sunshine of his prosperity is dimned, and, by circumstances which he cannot control, he sees his surest and most pleasing hopes ready to be overturned. And thus are we taught never to build too high,—not to think ourselves altogether secure in any situation. But, again, like the storm-beaten mariner,. just as he may be losing hope, and giving up all for lost,—just as he is beginning to think the gloom of adversity has thickened too much ever to be dispelled, he sees the darkness gradually removed—he sees prosperity again shoot forth her beam,—and again he finds himself in the prosperous course. And thus are we taught, never, in any circumstances, to let ourselves sink with despair, but, steadily to strive, even in the most unpromising case, and look forward to a better day. When we see the mariner guide his frail bark to the remotest corners of the world, and travel the very tract, and hit the very spot, he wishes, though he may never have travelled it before; we are taught how far science has advanced, and how much the ingenuity and industry of man can achieve, and we are stimulated by the noble works of those who have gone before us, not to repose in idleness, but to employ the faculties that have been given us,—the talent we possess,—in the purpose of utility. When we see him, on the approach of the storm, prepare his vessel in such a manner as enables it to ride secure, and gives his bosom peace, in a tempest which we would pronounce as almost certain to ingulph him, we are taught what exertion and proper management will do; we see how much, knowledge and steadiness and perseverance, will aid us, in unpropitious circumstances. When we see the dangerous lightning streaming along, and listen to the loud thunder as it peals, we are led to look up with fear and love to Him who has the lightning and the thunder in his hand, and who, when he wills, can quiet the storm.

Here, too, the feelings of pity are called into play; here the heart that can sympathize, and feel another's woes, finds a subject. When we see the poor seaman, for the safety of the vessel, obliged to expose himself in the very midst of the storm, obliged to get out in the dead of the night, perhaps only half clad, and climb the slippery shrouds, while the rain beats hard upon him, and the bitter blast benumbs and is like to carry him away,—when we see him night after night wanting his rest, and, perhaps, day after day without his regular food, surely, if there be any sympathy in us at all, it must be exercised here. It is only when we ourselves have witnessed the storm at sea, that we can rightly understand how laborious and disagreeable the life of the sailor often is; how many hardships and dangers he has to contend with. It is only when we have seen him engaged in his severe and perilous work, that we can know how very valuable he is, and how much we ought to prize him for the duty he has to perform, and how necessary it is that for his numerous toils he should be well rewarded.

Fortunately, even in the situation that we deem most unhappy, the man whose lot it is to be in it, accommodates himself to it, and finds a way of being happy. And the sailor, though we, who are only looking on, may suppose him to be leading a life scarcely worth having, has his happy seasons as well as others. When he has performed his hard duty, if he gets his can of grog, and his hard salt fare, ill as we may think of it, he makes a better feast, he has a healthier and far more grateful repast, than the man who has every luxury at his command; because he eats it with the appetite of health; he knows not the loathings of a pampered stomach. When, cold and drenched with rain, he lays down in his hammock, although he knows that in four hours he must again meet the storm, he enjoys a far sounder repose than he who lolls till the day is far advanced, on the bed of ease and idleness; because fatigue has prepared the body for the great restorer; indolence has not feverished and unnerved. So that were we to examine rightly, and come to just conclusions, we might, perhaps, find, that the sailor with all his privations, with seemingly scarcely a comfort at all, has really more enjoyment than those who seem to have every comfort, than those who may be contemning his situation. We know not what any thing is till we have actually reached it; and those ills, which, only contemplating at a distance, we think we could never exist under, if we come to be obliged to suffer them, change wonderfully in their aspect; and we not only find out that they are bearable, but that there is even some comfort with them! When fair weather comes, the sailor gets his ease, and never giving himself a thought about the blasts he may, before long, have to encounter, he makes himself happy.

There is also, when it blows pretty briskly, and the jumble of the sea is pretty considerable, something to annoy even the passenger. There is then, of course, a good deal of tumbling and tossing about; and frequently at the breakfast or dinner table some things occur, calculated, certainly, to vex, but perhaps, on the whole, to amuse more. In the first place, on the rough days, there is generally a shorter allowance than usual. Fewer dishes are served up, for a great many cannot well be taken care of. This, however, would not be much to complain of; but often the dish upon which we are placing almost our whole dependance, though we may think it pretty safely placed, follows, in an unlucky moment, a roll of the ship, ere any one has had time to look into it; and the chance is, that some unfortunate fellow to leeward, gets more of it than he is wanting. Or, if time has been allowed to help the contents about, as it is rather a ticklish thing, at such a time to eat and keep the seat too; sometimes, before the mess has been far proceeded in, some one or other to windward, more eager on his plate than his seat, loses his hold, and down he goes carrying others along with him; and those who, thinking they have got a safe birth to leeward, are looking out for no mishap, but solely engaged with what is in their hands, are all at once surprised, and very likely injured either in body or clothes, by half a dozen people, and half a dozen chairs, and half a dozen plates of soup, going down upon them. And as a man in jeopardy will catch at a straw, when those who have lost their hold find themselves moving in a downward direction, although there is nothing in the way to seize upon, but a knife, or a fork, or a tea-pot, or a tumbler, or a wine decanter, even these they will grasp at in their eagerness to save themselves, and in their attempts to prevent, sometimes aggravate, the misfortune; some one, perhaps, gets a touch he would rather dispense with. And sometimes, to put a good finish to the whole, while the fallen are busy recovering themselves, and those who are still fast are engaged either pitying, or making light, of, their distress, another big and unexpected roll comes, which either sends them down on their fallen brethren, or walks the remainder of the eatables off the table, and the only comfort left is, to look forward with hope to the next meal.

> "Sweet are the uses of adversity."

And

> "He cannot be a perfect man
> That is not tried and tutored in the world."

"But I would you did but hear the piteous cry of the poor souls; sometimes to see them, and then not to see them; and now the ship boring the moon with her main-mast, and anon swallowed with yeast and froth, as you'd thrust a cork into a hogshead. I would you did but see how, *when the soup and the big carving knife lighted upon him*, he cried to me for help, and said his name was *Antigonus*, a nobleman; and how, as the poor souls roared, the sea mocked them."

Another very gratifying scene at sea is, when, on the Sabbath, the men are all mustered, clean and in their best apparel, to join in common worship. There is something exceedingly impressive in hearing the word of God preached and read in such a place. There is something that lifts the feelings high, and makes sure of carrying them away from all common affairs, that lights up the flame of devotion, and infuses the spirit of seriousness and kindliness into the heart, even into the heart that at other times may sadly want religious feeling. The clear blue heavens above, the far-spread ocean on every side, the gentle propitious gale urging the vessel on her course, the attentive crew sitting round, the man of God in the midst, preaching zealously the instructive lesson, or offering up the solemn prayer, give an interest and a solemnity to the scene which is truly impressive. It seems, in such a place, as if we were removed in some

E

degree from the world, as if we were brought a little nearer to the Creator; and the mind cannot help being drawn towards Him who, presiding over all, sends, as he sees meet, the favourable gale to urge peacefully on, or the mighty tempest to destroy. As the commands of that One are set before us, as the duty we owe to him is explained, who is it that does not listen. As the warm prayer is offered up for a prosperous voyage—for the welfare and prosperity of our native land—for health and blessing upon friends far distant,—who is it that does not answer in sincerity of heart, amen!

Thus far our Saviour's tender care,
  Has brought us safely o'er the deep,
And charg'd the winds and waves to spare
  A few, the meanest of his sheep.

Oh! let our souls with praise record
  The thousand mercies we enjoy;
Beneath the safeguard of our Lord,
  Kept as the apple of his eye.

God of our hope! to thee we bow,
  Thou art our refuge in distress;
The husband of the widow thou;
  The father of the fatherless.

In midst of dangers, fears, and deaths,
Thy goodness we'll adore;
We'll praise thee for thy mercies past,
And humbly hope for more.

There exists, I think, in almost every mind, a natural inclination to look back upon what was once enjoyed; to ponder upon things past, upon objects removed. And, at sea, to him who possesses a very reflecting mind, there cannot be a more delightful employment, than occasionally to withdraw from the bustle, and in the retirement of the cabin, or on the lone deck at night, give the thoughts to happy remembrance. It is wrong to enter upon reflection that can only cast a gloom over the feelings, and it may be thought that the reflection of the voyager, who is far away from his dearest objects, would just be of this kind; and, therefore, the less that he indulges in meditation, the better. But this I know is very far from the truth. I have felt what the sensations are, when we call to mind those we love, though they be far from us, and I pronounce them just to be those that will elevate, and give relief, to the mind. Oh! there is a feeling inspired, as

E 2

we allow the imagination to stretch to the home of our affections, and bring before us what we so long were associated with there, which restores vigour to the eye that sorrow has bedimned, and the pulse of joy to the heart that has long beat in sadness. For the time, we, as it were, hold converse with those we meditate on; and though it be but an intercourse by imagination; though it be but a feast in idea, it is indeed a great one. As we know not of any change that has taken place in our absence, although fears may intrude, yet hope, ever more powerful, dispels them, and we picture those we left in the same happy state in which we left them; and, as thus viewing them, we heave the sigh of affection, and breathe forth the prayer for happiness long to continue with them, we gain a serenity and firmness of soul much to be desired. The sordid few who possess scarcely a tender feeling, or him that never strayed from the scenes, and tore asunder the attachments, of his infancy, may not be able to appreciate this retrospect of fancy, but ever will it be regarded by the feeling wanderer as his great support—as one of his greatest comforts.

And, after all, it sometimes only requires just a little stretch of the fancy, to make us, even in the middle of the Atlantic, almost believe ourselves on terra firma,—in some rural situation. Frequently, in that very place, have I been awakened in the morning by the crowing of the cock; and scarcely had I my eyes opened when I was saluted by the squeaking of the pig; then followed the bleating of the sheep, and the bellowing of the cow, the barking of the dog, and the soundings of all the different poultry; so that really, had it not been for the continual cracking of the beams and bulk-heads around me, and the unsteadiness of my footing when I got out of bed, and the cables and spun-yarns and marline-spikes that were in view from my cabin-door, I should have had little difficulty in supposing myself very near indeed, to a barn-yard. Yet sure as I was of my pent up situation, the chorus I have mentioned has often filled my mind with the most pleasing recollections, and often, ere I rose from my pillow, have I visited, in imagination, many a dear and lovely landscape, and traversed, in all the fondness and eagerness of reality, the mountains and valleys which saw my childhood.

In short, every life has its pleasures and its pains. All depends upon the disposition; upon how we occupy, and allow the feelings to run; upon how we strive to accommodate ourselves to existing circumstances. It would certainly be maintaining what is not true, were I to aver, that a life at sea is as comfortable and desirable as a life on shore. Most people, I believe, and I confess myself to be one of the number, would prefer, on the whole, the latter. But what I mean to maintain is, that even the former is not without its comforts and pleasures; and that he who follows it, either from choice or necessity, if he thinks and acts properly, will not be without enjoyment.

CHAPTER III.

THE APPROACH TO THE INDIAN SHORE.

Th' impatient wish that never feels repose;
Desire that with perpetual current flows;
The fluctuating pangs of hope and fear;
Joy distant still, and sorrow ever near!

FALCONER.

THERE is an anxiety—a restlessness—in the bosom of man, which is continually presenting something new to his imagination, and urging him on to some new project. He is seeking after contentment, after comfort; and he fixes upon a point where he makes sure it is to be found, and promises himself that when he gets to this point he will have gained the summit of his wishes. There he will rest free from desire, for there all that he wants will be supplied. But, when he has reached it, he finds not what he counted on; something is still wanting; some feeling is still unsatisfied, and ere he can gain altogether what he is in quest of, he must press to another point a little further on. Yet, let him reach even this, and still he has not all. His views change, and new objects appear, and new wants are found out the farther on he goes. Just as one wish is satisfied another takes its place; as one difficulty is overcome another starts up; the object flits just as he speeds. There is, indeed, no boundary to his desires, no point at which he can gain contentment; and thus does he go on projecting and plodding; thus does one wish follow close upon another, and ever is his bosom in restless eagerness, until death, perhaps very unexpectedly, steps in, and carries him away from all his hopes and all his ambitions.

Does he not after fairy shadows run?
Follows he not some wild illusive dream?
Like children, that would catch the radiant sun,
Grasping its image in the glittering stream!

And this spirit of anxiety is well exemplified

in an Indian voyage. Immediately the ship has started from her port, every one is wishing for a certain wind, but no sooner has this wind come and satisfied the wish, than they have something else to ask; they would like to be a certain length on—in, some particular latitude. Well, by and by they get to this latitude, but still there is just as little appearance of satisfaction as before. Some other point is immediately marked out; if they were just half way, they would be pleased. But, when the half way has been fully measured, and we expect that all anxiety is ended, still we find it as before. Indeed, the anxiety increases just as the distance diminishes; and, when the vessel has nearly run her course, and the land is supposed not to be far off, then is it seen at its greatest height, and the eagerness then evinced by almost every one is truly astonishing. Then those who never ventured from sure footing before, are seen high upon the mast, eagerly looking out for the expected object; and then there is such calculating and prophesying; and should the wind veer in the least from the favourable point, oh! there is such grieving and lamenting. Frequently is the land seen long before it is in sight, and its fresh fragrance felt at a most unconscionable distance. And a bird which is known never to stray far from the land, or a little piece of sea-weed, because it is likely to have come from the shore, or a butterfly, which must have travelled from the same quarter, is then hailed with as much joy as if it were the Governor of the Indies himself come to receive and welcome them.

Now, I think I am right when I say, that this spirit of restlessness, although it appears to be natural, to be congenial with the feelings, when it exists in a very great degree, is rather an unlucky ingredient in the disposition. By giving ourselves so much thought, and looking with so much anxiety upon things that are before, we notice less, and lose much of the enjoyment we might draw from, those that are really present. It seems as if this continual stepping of the mind from one object to another, by keeping it always on the stretch, gives it no time to dwell upon, or really to find out, any single circumstance of gratification; and thus, in its eagerness to reach the very summit of happiness, hardly to taste of happiness at all. It is certainly necessary to possess the spirit in some degree. As we are so liable, in every situation, to have our hopes so often disappointed, if there were not something to stimulate and urge us on—if we could not fix them again upon some object before, we would sink into a miserable state of indolence and despondency. It is well for us, no doubt, that when the mind loses one hold it can readily seize upon another; that upon the ruins of one prospect it can instantly build up perhaps a better; for thus it is kept strong and encouraged to go on in the pursuit of happiness. It is our hopes and fears that give us life, that keep a heartless apathy away from us. But when the anxiety increases along with the success; when, after a wish has been satisfied, a thing obtained, it is no more valued, but something else is looked for with still greater avidity; in a word, when desire follows desire, and there is no gaining contentment even under favourable circumstances, and the mind is ever kept in a racking, unsatisfiable state of expectation, surely it is a spirit which cannot be advantageous, and which should not be encouraged. And therefore have I long thought, that he who endeavours to make himself happy in all situations, and under all circumstances; who, while he looks to the future with good hopes, appreciates and enjoys the present, follows the wisest plan. Therefore am I fully convinced, that he who, in a long voyage, keeps himself much the same at all times; who frets not with the wind when it opposes his progress, but quietly waits till a change comes; who, although he would be very well pleased could the voyage be got over in half the usual time, will not grieve very much though it should be rather longer than the usual time, is by far the happiest man. And what is more, he loses nothing by his contentment; for he gets to the end of the voyage just at the same time with those who have all along been warring with the winds, and still more with themselves than the winds.

And in this, as in every thing else, when we are too eager, we are apt to miss and be deceived.

> *Hamlet.* Do you see yonder cloud that's almost in shape like a camel?
> *Polonius.* By the mass, and 'tis like a camel, indeed.
> *Ham.* Methinks it is like a weasel.
> *Pol.* It is backed like a weasel.
> *Ham.* Or like a whale?
> *Pol.* Very like a whale.

I have known one started from his bed, at rather an earlier hour than he was in the habit of getting up at, to see land, which, after all his trouble of coming upon deck, he found to be made up of a cloud. I recollect, when we were drawing to the Indian coast, and expecting very soon to see the island of Ceylon, of more than half a dozen people standing for some time one night, inhaling, with wondrous pleasure, the "spicy gale," as it was wafted from the orange and cinnamon groves of that lovely island, and making themselves sure that we could not be many miles from it; when it was afterwards found, that we were nearly as many degrees, and that the whole fragrance had proceeded from a little Friar's balsam on one of the party's fingers. I have seen nearly the whole of a ship's passengers drawn from their different employments, and sent in no small haste upon the poop, to see a bottle with something white about its neck, or a piece of light white paper, shaped somewhat like a butterfly, or even a plain bit of old rope, which some waggish chap had thrown out at the bow. And therefore, it is highly necessary, at such a time, to have all our eyes and senses about us, that we may see the land just as soon as it appears, and smell the scented wind just at the time it reaches us.

Yet, for all this eagerness to reach the destined shore, the mind is not altogether free from uneasy sensations; there are reflections of a depressing kind apt to intrude as we draw to it, and the heart that beats high with expectation, while the land is yet unseen, may lose all its buoyancy when it is seen. Now that we have really arrived at far-famed India, we feel, indeed, that we have left, at a great distance, the land where all our affections remain; and the happiness we experience in again seeing the wooded coast, and the far-stretching green-mantled hills, is much over-clouded by the reflection, that it is the coast and the country of strangers. This may be the country which some dear relation, or intimate friend, visited like us, and went to it to return no more. This is the country so inhospitable to the European; where the frail body is so apt to become the victim of disease; where the strong man must put no faith in his strength; where the sweeping epidemic, stretching far and wide, exerts its dreadful influence over all, and day after day gathering new strength, carries on each succeeding day increasing numbers to the grave. And thus, we think, it may be with us; in the general wreck may we sink unheeded, unmourned. In a foreign land, far from all that we love, from that which can alleviate so much the loneliness of the death-bed; which can cheer so much the expiring heart, and brighten so much the sad gloom of the dying hour, may our eyes be closed.

And even if we escape such a fate, although, after a time, we may go back in safety to the land of our fathers, yet, as many years must roll past ere that period comes, consequently many changes must have taken place, and many of those ties that now bind us to it, shall ere then be unloosed for ever. Some of those that are dearest to us will, in all probability, have gone down to the grave; for the old must droop and die, in the common course of nature, and even the young are not secure from an early stroke. Long absence may have deadened and weaned the affection; and the joy-beaming look, and the warm-heaving bosom, which our presence ever inspired, and which stirred up in our breasts the grateful responsive glow, may be gone for ever. The companions, too, of our childhood and youth shall be scattered and lost. The few we may find we will scarcely recognise as our early companions; for time will have been busy with its changes, and new associations will have been formed. Scarcely a lineament of the disposition which once existed may remain, and both the joys of former days, and the companions who shared in them, may no longer be remembered. We shall, indeed, just be alone in the midst of strangers; and the remainder of our years must we wear out, as it were, in solitude, deprived of those comforts which give such beams of consolation to the afternoon of life, which cheer and support man as he journeys to the grave.

Better then, we naturally exclaim, would it have been for us that we had never wandered so far; if such are to be the consequences, surely we have deceived ourselves—surely we have judged most unwisely in ever straying from our home. What are all the riches we can accumulate, or all the knowledge we can acquire, or even all the honours we can attain, by our wandering, if the probability is, that ere we have had time to enjoy them, we will be taken from them? What are all the wonders a foreign country can present, or all the luxuries it can give, if we are to buy them at the expense of far truer, though perhaps less glaring, enjoyments? What, in a word, is every thing else, if we are to resign all that ever interested our bosoms—if we are to give up that happiness which is above every other, and which, having once given up, we shall probably never taste again? Happier by far, is the poor peasant, who, though earning his bread by the hardest labour, yet eats it in his happy home, in his native healthy atmosphere. Happier, surely, is he who, with just enough to keep him above the world, has his kindred and kind companions around him, and enjoys the little which Heaven has bestowed on him, in the society of those he loves, and in the place that infancy has naturalized and made pleasing to him. Ah! happy are they in whose hearts contentment dwells; who, satisfied with humble joys, have never given place to a wandering or an aspiring thought. And well would it have been for us, had we checked, as it came forth, every wandering desire, and never have allowed either curiosity, or the hope of wealth or distinction, or any other enticement whatever, to draw us from our first and happiest state. It would have saved us many a pang, had we moderated our desires and kept ourselves at peace in the situation we were placed in; had we held fast the humble blessings we possessed, which, little as we valued them then, we now find to be inestimable, instead of wandering in search of those which it is probable we may never find, and which, even if we do find them, may not give us the satisfaction we counted on.

These are the reflections which are apt to arise as we draw to the strange land; and certainly they are, in so far, true. The joys of home are by far the sweetest; and he with but a little in the bosom of his family, if he feels as he ought, will have far more comfort than the man of thousands, in the land of strangers. But though the reflections be natural, though they have even truth to recommend them, they ought not to be much indulged; for it would suit very ill indeed if they were received and acted on. Circumstances will occur, which render it absolutely necessary that we should travel far, and leave all behind, and leave it perhaps for ever; and therefore it is well that all do not possess the same quiet and unenterprising disposition; that there are some who will never give place, or at least never allow themselves to be overcome by such reflections as these. When we reflect that the bright aspect which society now presents, the immense progress which science has made, the knowledge, the prosperity, the beauty of nations in general, are the fruits of the wandering and the enterprising mind, we see how good it is that such minds have been. When we look at the page of history, and see it abounding and sparkling with the noble deeds of the wanderer,—when we see what has been achieved in the cause of Freedom, and in the cause of Philanthropy, by the efforts of those who forsook all, and sacrificed much, and entered upon labours and hardships the most distressing, and fronted even dangers they were likely to perish in; we see the value of the daring spirit; we see how happy it is that such men have been; how necessary that such should be. Nay, when we look even into the common and private scenes of life, we very soon discover the frequent necessity and expediency of the disposition. When we look into the concerns just of the domestic circle, we are not long in finding out, that it is indeed a fortunate thing, that while some prefer the quiet pleasures of home to the more dangerous, though more enticing ones, abroad, and sit contented, others have stronger and more towering feelings, and hesitate not, if it seem necessary, to embrace the latter.

And therefore he who is approaching a foreign shore, if he does wisely, will endeavour to suppress as much as possible these feelings of despondency. They cannot afford him the least relief; they can only lay an additional load upon the mind which has perhaps already enough to burden it. He may not be able to forget altogether his sacrifices; he may not be able to forget altogether his dangers; but he should strive to eye the former without much regret—the latter without much fear. It is not indeed necessary that he should forbear looking back, neither will he be the worse for contemplating and examining what is before: but it is necessary that he views the past with firmness, and the future with undauntedness. Certainly let him meditate, but let him meditate with proper feelings; perhaps his meditation might be this:—I have exchanged, it is true, a comparatively secure and happy state for one where I am exposed to numerous dangers, and where probably no happiness may be found; but it seemed necessary that the exchange should be made, and therefore I ought not to despond, but stand against my difficulties, with all my efforts, and make myself just as happy, as contented, as I can. There are many dangers threatening, yet I have much in my own power; much depends on my own conduct: and though I cannot make sure of steering clear of them, still I can do much to avoid them. The more eagerly I labour the sooner will my labour be finished, and the sooner I will be able to return to the place I have left. And if, at my return, I find the same kind friends ready to receive me, if I again meet the warm embraces I so long enjoyed, then I shall be able to appreciate them the more for having once left them. Yet, hard as it is, if those I love should be taken away, and I am left, as it were, alone in this vale of life, still I will not be without comfort; for the very associations of the scene will go far to give me happiness. Memory will be ever busy in supplying food for the lone mind. The place I am in, though it may continually tell me that my greatest friends, and my greatest comforts are for ever departed, yet will it thus as continually keep them fresh in my remembrance, and enable me, fled as they are, to draw happiness from them till the hour that I drop into the grave. And, even though it be decreed I am never to visit my home again, and never to taste even of these ideal joys; though it be doomed I am to die in the land of strangers, still am I content; for so be it I die well, it matters little where—

Of chance or change O let not man complain,
Else shall he never never cease to wail;
For, from the imperial dome, to where the swain
Rears the lone cottage in the silent dale,
All feel the assault of Fortune's fickle gale:
———————— Let those deplore their doom,
Whose hope still grovels in this dark sojourn;
But lofty souls, who look beyond the tomb,
Can smile at fate, and wonder how they mourn.

It has been beautifully and truly said

Our affections
Must have a rest; and sorrow, when secluded
Grows strong in weakness. Pen the body up
In solitary durance, and, in time,
The human soul will fix its fancy
E'en on some peg, stuck in the prison's wall,
And sigh to quit it.

The mind accustomed even to a disagreeable state comes in time to imbibe an affection for it. It seems to be a law, that whatever we have been long connected with, although, at the time the connexion was begun we disliked and did not choose it; although all along we may be discontented and continually wishing the connexion broke; yet, when the time comes that we have it in our power to break it, in doing so we feel some reluctance. Although we never dreamed of it before, at the time that we are about to break the tie which binds us to the hated object, we find that somewhat of our hatred has fled, that enough of affection exists to draw forth a sigh as we break it. And more—the affection is not the feeling of a moment; it remains—and very probably increases, after the bond is loosed; for ever after, the disagreeable state may be looked back on with kindness; nay, it may even be wished for. The man who for many years has been kept plodding continually in the walks of business, considers it a most laborious, unsettled, and uncomfortable life; and his greatest hope is—that by-and-by he will be able to give it up, and get into some snug spot of ease and retirement. But when he has accomplished his wish, he finds not the enjoyment he looked for, and ill as he liked it when obliged to be in it, he sighs again for the bustling world. When the unhappy prisoner, who, for half a century never had crossed the threshold of his lone cell, was allowed, on the ruin of despotism, to go back into the world, he found not any comfort in it, and he returned and demanded his dungeon and his chains that he might again have happiness.

And in an Indian voyage we have also an illustration of the feeling. Although at the beginning the adventurer finds himself miserably situated; although to the very end he may look on his situation as a most tedious and uncomfortable one; although every day there may be stronger wishes for the termination of the voyage—and, at last, indeed scarcely any thing else seems thought of; yet when the period of his release is drawing near, he finds he can look with a little less dislike on his situation; when his strongest wishes are just about to be fulfilled, he finds they have lost a little of their strength; his eagerness to get rid of his unhappy birth has abated considerably, and even something like affection for it appears. When he is stepping from the ship, the scene of all his miseries, the place of his bitterest execrations; when he is exchanging the little cabin and the tumbling cot, for the secure bed and the spacious bed-room—he execrates none; he dwells on his hours of happiness in it, rather than on his days of sorrow. Something has carried away every unkindly feeling, and regret, ay fond regret, as he eyes it for the last time, stirs and warms his bosom. And ever after may he look back with fondness on his days of voyaging. When, long afterwards, he looks to the time when he deemed himself most uncomfortable—to the time when he deemed himself away from the world and all enjoyment, he may be inclined to say, that though he deemed them so then, these were not his unhappiest moments,—that then he had indeed more enjoyment than he has, sometimes, had since in the bustling world; and he may even wish again for another trip upon the waves.

And again, in this feeling we have a beautiful proof how much that restlessness and changeableness which exists in man, stands in the way of his comfort. It shews us that the eagerness which is ever keeping him in anxiety, is also ever leading him into error; that in looking so earnestly for what he expects, he is apt to neglect what he really has; that sometimes when he ought not, he frets and discontents himself, and finds hardly any enjoyment in the situation he is in, merely because he has fixed his desires upon what he imagines a better. It shews us that after we resign, or just when we are on the eve of resigning, what we deem an unhappy state, we may find out that there is indeed happiness in it,—that it was truly worth our prizing. And shewing us this, it teaches us, that nothing we are obliged to engage in, we should consider too hard, or rashly condemn and throw up; that whatever our state is, we ought to endeavour to be content, for it may be better than we imagine; it may be better than even another we are desiring. It teaches us that even all the privations and annoyances of an Indian voyage should not make us complain very much. And lastly, seeing that the feeling is apt to exist under all circumstances,—that the mind habituated to any thing, whether good or bad, has a reluctance and a difficulty in withdrawing itself from it; we see how nicely we ought to choose,—how necessary it is that we strenuously avoid making any attachment we would by-and-by wish to have broken; and how quickly we should set about breaking it, if such an attachment has been made.

CHAPTER IV.

THE ARRIVAL.

> ———————— Where'er we roam
> Our first, best country, ever is at home.
> And yet, perhaps, if countries we compare,
> And estimate the blessings which they share,
> Tho' patriots flatter, still will wisdom find
> An equal portion dealt to all mankind.
>
> GOLDSMITH.

At dawn of day on the 21st of July, after a passage of just four months from England, we anchored safe in Madras Roads. The preceding night even the passenger was found keeping watch. Some, I believe, did not go to bed at all, but remained upon deck, looking out with as much anxiety as the commander himself for the light which was to warn us of our approach to Madras; and most of those who did go, went only to remain, awake, until they should hear the glad call 'all hands up to bring the ship to anchor;' so anxiously did every heart beat as we were drawing to the port.

I have formerly spoken rather against an over-great anxiety at sea, to get forward; but I can well excuse, nay, I can almost commend, a little eagerness at such a time as this. The thought that after all our dangers and anxieties we are now getting into the harbour of safety: the inclination which all have, more or less, of seeing the end of what they are engaged in; the curiosity we possess to view the far-distant and far-famed place; the place where, perhaps, we are destined to spend many of our years; stir up in the bosom a degree of anxiety which it is difficult, indeed scarcely possible, for any one to restrain. Nor is it right perhaps, as I have said, that it should be restrained. For it is not the fretful anxiety of disappointment. It is not that racking of the mind caused by discontentment; that anxiety which exists in an over-eager, ever-asking spirit. It is the gentler anxiousness of natural desire,—the more grateful buoyancy of gratitude and hope and curiosity; and as it undoubtedly serves to occupy the feelings and keep away despondency, it serves a very necessary purpose.

When the morning had advanced a little, and the town with the surrounding country could be pretty plainly distinguished, every eye was bent towards them, and every heart was the seat of many varied sensations. The sun rose upon a serene sky, and, as emerging from the ocean, he threw his bright beams on the majestic white building and the green scenery surrounding, objects we had not seen for long before, the scene presented was indeed a gratifying one. It recalled to us objects like these, better known and higher prized. It reminded us of some of our dearest scenes, of some of our happiest seasons. And though it sent regret into the heart, yet joy went with it; the fond, accompanied the regretful feeling, and took away its bitterness.

When the scene had been long viewed, the desire to visit naturally followed; and by-and-by, the half-clad sleepless gazers disappeared, to make their different preparations, and get equipped for landing. Very soon after we anchored the natives came on board in great abundance, eagerly offering their services both to take us on shore and attend us while there. And in course of the forenoon, as we got in readiness, we got into a clumsy Indian boat, manned by a very noisy strange-looking set of chaps, and proceeded to the land.

To those who have been a length of time at sea, it is indeed a gratifying thing to get the foot once more upon dry land. The shore, after all, is a more natural place for us than the rough sea; and there are few, I believe, however much they may be attached to the seaman's life, but are glad to exchange it now and then for the landsman's. And it is only those who have gone a long voyage, that can have a right idea of the very grateful feeling produced when we touch our mother-earth again. The gravest, I am sure, will hardly be able to keep from taking a jump of joy; and, at such a place as Madras, a jump of this kind is an excellent thing, that is to say, if we jump in the proper direction; for if we do not look out pretty sharply, the big surge which rolls there may teach us, too late, that we are not quite upon dry land even when we imagine it. We are inclined, after getting fairly on the landing-place, to stand and consider a little—to make sure that we are really once more upon terra firma: for indeed there is a little difficulty, after having been so long tossed about, and had our course so much circumscribed, in assuring ourselves that we really have a firm footing again, and a clear wide range of country around us. But, at Madras, there is very little time allowed for consideration; for immediately the new-comer gets upon the beach, he is surrounded by an immense flock of, to him, very odd-looking fellows, who astound him so much with their loud, uncouth, and continual gabble, and in their eagerness to get at him and offer their services, press so roughly upon him, that whenever he recovers a little from the astonishment which such a strange and unexpected salutation is sure to produce, he is glad to make his way out of the group as quickly as possible, and get to some place of quietness and safety.

At Madras, some of our passengers remained. There are few who have not experienced, at some time or other, the pangs of parting. There are few who have not had to bid adieu to some dear relation, or long-tried friend, and experience the grief which such a parting calls forth. But it is not such a parting that I have now to allude to; it is not the bidding adieu to one that we are bound to by the warm ties of relationship, or one that we have long and intimately known; but it is the parting with those whom we have only been with for a little time, and with whom acquaintanceship is yet young.

Most people, I think, have been in situations where, for perhaps a few days, they were kept pretty closely with strangers; and though no particular intimacy seemed to be formed, yet, when the time came that they were all to separate, and all to go different ways, probably never in this world to meet each other again, there was a depression of spirits produced, a degree of me-

G 2

lancholy was cast over the mind, which they did not look for. And when, instead of days, we have been for months, as in an Indian voyage, in close association with strangers, it is natural to expect that the melancholy feeling should be felt still more: and we find it to be so. When the vessel has reached the port where some one is to take leave, although up to that time, it might not have been imagined that any attachment had taken place, it is then felt that intercourse has indeed been productive of affection, and each heart feels a pang when the "farewell" is given; and those who remain to go farther on with the vessel, feel a wonderful want in the little society. He that has a spark of tenderness in him at all, will find his heart expand with the warm glow of affection, and the tear will be ready to burst from his eye, as he grasps the out-stretched hand, and bids adieu to the man he is, in all probability, never to meet with again, even though he may not have known that such a man existed till he entered upon the voyage. He that, when thus taking a last farewell, would not press even his enemy to his bosom with all the warmth of a forgiving and affectionate soul, he lacks much of kindly and honourable feeling, and deserves not to know the value, the comforts, of friendship. It is only what is due from one man to another; it is nothing more than our common nature—our mutual relationship demands.

> " There be, perhaps, who barren hearts avow,
> Cold as the rocks on Torneo's hoary brow;
> There be, whose loveless wisdom never fail'd,
> In self-adoring pride securely mail'd;
> But, triumph not, ye peace-enamour'd few;
> Fire, nature, genius, never dwelt in you!
> For you no fancy consecrates the scene
> Where rapture utter'd vows, and wept between;
> 'Tis yours, unmov'd, to sever and to meet;
> No pledge is sacred, and no home is sweet!"

And here again we have an illustration of that feeling alluded to in the preceding chapter, which teaches us that affection may exist when we deem it does not; that it may be just when we are resigning an object we find out what it is. And here again we are taught to prize what we possess; to seek out the affection, and draw the comfort from it which it can give, ere the time it is to disappear—to value the friend while yet we have him.

On the 6th of August, we again put to sea; and on Friday, the 17th, we reached Calcutta, and finished the voyage.

He that visits the Indian shore, cannot fail to be struck almost as much with the situation the English hold there, as with the natives themselves. It is, indeed, astonishing, to see with what fear and respect they are there looked up to, how far they seem to stand above, and how much authority they exercise over, the numerous natives around them. In the bosom of the Britainer it calls up no inconsiderable degree of wonder, as well as gratification, when he sees such a vast and immensely peopled territory, such a fertile extent of country, completely under the dominion of his countrymen, and giving up to them almost the whole of its great produce. It gives him an exalted opinion, indeed, of what Britain can achieve; of how much courage she can muster up, and what strength she can put forth, when he contemplates her, insignificant as she appears, holding so completely in her grasp, such a powerful and far-distant country. And higher must his opinion rise, when he remembers, that ere she could win it to herself, she had not only to contend with and vanquish the possessors of the country themselves, but also with others, seemingly alone more powerful than she, who were also contending for the prize. When we see the flag of England streaming almost exclusively in the port, and the wealth of the interior almost exclusively carried away to the English shore, and the mightiest chief of the land bowing to the command of an English governor, it tells us, most convincingly, that there is vigour in the British arm, and valour in the British heart. It tells us, that whatever the giddy few may cry out—that whatever discontent and clamour England may have heard, and however insignificant she may seem, there is a mighty spirit in her bosom, and wisdom with it; that in the day of need, every division can cease, and there can be a general unity and a general rallying of her sons; that when her glory is to be maintained or added to, her spirit is ready to burst forth—her wisdom is ready to be shewn.

And much is our gratification heightened when we see the power she possesses exercised with clemency; the laws she has enacted wise, and exercised with justice. The limits of right and wrong she has made known to the native as well as to the European, and both stand equally amenable to the law for any act of transgression, and equally under its protection, so long as they obey its dictates. The chain of slavery does not exist, and the arm of tyranny has no power. The poorest man is free; and while he becomes liable himself to punishment, if he is guilty of any fault in what he binds himself to perform, he has a power to apply to, and a power that will hearken and give justice in his cause, if he finds himself aggrieved. It is to be hoped that few, even were there no such strict restraint, would tyrannize over the poor Indian in his grasp: but it is good such a restraint does exist, for it ensures the conduct which duty dictates, and it shews that in the ruling power there is justice and humanity. When all see with what strictness the laws are executed, they are necessitated, whatever may be their inclinations, to fear and pay respect to them. When the native sees punishment follow, with due force, upon the white man as well as upon him, it gives him an exalted opinion of the arm that is over him, and attaches him to it; when he finds us shewing him kindness so long as he acts as he ought, he is naturally stimulated to conduct himself with propriety; when he sees us doing much both for himself and his country—when he feels his situation not made worse, but better, under the hands of his vanquishers—he begins to find out the wisdom and the virtues of our customs, and he hesitates not to lend his ear to our instructions, and imbibe our better notions; and he comes to count the day indeed a happy one, when a more civilized, though strange and intruding, people, first visited him.

There are however a few that would look upon the native as a different kind of being altogether from themselves, and as neither deserving of the same usage nor the same justice. Seeing him in nature's uncouthest state, wanting altogether the cultivated manner, and wanting, at least in their idea, the knowledge and vigour of mind which they themselves possess, they regard him as a being far beneath them, and treat him with a contempt and a severity which he does not deserve. He cannot, perhaps, accomplish what they want him to do, just so expeditiously or intelligently as they wish, or as they could do themselves, and therefore they bring the charge of great stupidity against him. He is now and then guilty of a fault—nay, we cannot deny that sometimes he is guilty of an aggravated crime—and therefore do they bring against the whole race the charge of treachery and dishonesty, and maintain they are deserving of no sympathy—of no kindness. But, if we just consider a moment, we cannot fail to see how unfair and inhuman such charges are. Are we to expect that the poor Indian, brought forth and brought up in the very bosom of ignorance, without a single opportunity of cultivating the intellect, or of catching a single notion, save the notions which a wretched superstition has set before him, is to possess an intelligence or an acuteness any thing like ours? or are we to expect to find him without fault?—we who are more civilized—we who plume ourselves so much on our knowledge and civilization—have also to own with many a crime. We cannot deny that dishonesty and treachery, and even crimes of a deeper die, are sometimes found amongst us; and therefore should we neither wonder nor condemn very much, if we find the Indian, tutored as he is in the dark school of superstition, and taught perhaps from his infancy not to look with much affection on the white man, transgressing frequently. His stupidity, it is plain, is just a natural consequence of his ignorance; and the man who acts as justice and humanity dictates, will not treat him with harshness on account of it; he will endeavour to remove the darkness of his mind, and, as he accomplishes this, the stupidity will disappear: he will find even the Indian's mind capable of cultivation. His dishonesty and treachery must proceed in a considerable degree from the same cause. It proceeds so far too, we are afraid, from the example of the white man himself; for, it cannot be denied, that sometimes those who cry out so much against him, and whose duty it is to set before him the example of regularity and equity, do not give him the pattern they ought. And it proceeds, I hesitate not to say, in not a few instances, from the treatment he receives. There never was a heart reclaimed, or won over yet by ferocity. It is much easier to woo by kindness, than to force by fierceness. It is kindness which gains the affection, and which keeps it after it has been gained. It may be possible by ferocity to make a man fear, but it is impossible to make him love. And the Indian, though he may crouch under the frown, while he labours at the work of the man who, lorded over him, treats him harshly, will never have an affection for this man, or a desire to do his work diligently; and no wonder if he be ready, when an opportunity offers, to take his advantage. And thus it is easy to see, that we ourselves are in some degree to blame for his dishonesty, and that his dishonesty is not likely ever to be removed, until we be scrupulous in setting before him a strictly honourable example, and be ready, when we find them out, to praise and reward his fidelity and services.

I do not mean to exculpate the Indian altogether. I know, whatever may be the cause or causes, that there is enough of treachery in him to make it absolutely necessary that he be kept well under restraint. I allow that sometimes he shews, even when he ought not to shew it, ingratitude and infidelity; and that we would require to be on our guard, and not give him all our faith, until time has shewn him worthy of it. But what I maintain is, that there are virtues in the Indian's heart, and powers in his mind, if they be properly sought after; and that while we treat him with that strictness which will shew him he is not to be allowed to usurp, we also treat him with that kindness which will shew him that we are not inclined to trample. There have been examples enough to prove that his mind is susceptible both of improvement and affection. No one can have been long amongst his kind without discovering, in many, an immense degree both of acuteness and industry. The cases are many, where those placed in the situation of improvement, have shewn great powers: nor are they fewer, when placed in the important, but kind, situation of trust, they have shewn great fidelity and great affection. And therefore is it that I maintain, that we ought to look with pity and kindness, rather than with contempt and harshness upon the being whose lot has been cast in a situation which prevents him from having feelings like ours; and that we will gain more, and be deserving of a better name, and fulfil just what duty requires of us, by complacently endeavouring to do away with his darkness. and root out his evils, than by looking on him as a being beneath us, and domineering over him, and quarrelling with his faults and failings without making an attempt to remove them.

" I venerate the man, whose heart is warm,
Whose hands are pure.
I would not enter on my list of friends,
(Tho' grac'd with polish'd manners and fine sense,

Yet wanting sensibility,) the man
Who needlessly sets foot upon a worm.
An inadvertent step may crush the snail,
That crawls at evening in the public path;
But he that has humanity, forewarn'd
Will tread aside and let the reptile live."

It has already been hinted at, in the second chapter, that in rating the happiness of a people, or an individual, if we rate upon the principal that what gives happiness to us would also give it to them, we will often rate erroneously; that to estimate any state aright, circumstances must be properly taken into account. And we have a noble example of it in the case of the Indian. Were we to measure his happiness by our ideas, we would scarcely conceive him to have any happiness at all. With only a little mud hut to screen him from the burning heat and deluging rain, and with only a little rice to subsist on; debarred, seemingly, from every source whence comfort flows, we are apt to look upon his state as a wretched and unhappy one. We would not suppose that comfort could exist in such a condition. But judging thus we would judge erroneously. For if contentment is comfort, and surely it is, it exists there. If a bosom without anxiety is a happy one, and I think it is, the Indian has it. Give him but his rude hut to repose in, and a little rice to support him, and his hubble-bubble, or smoking pipe, and he has not another wish; he has all that he wants; he has all that ever he knew; he has enough to make him happy. There seems to be more peace in his heart; perhaps more real enjoyment in his state, than in that of the man who is lorded over him, and who has it in his power to draw his comfort from a thousand sources.

And when I contemplate the state of the Indian, when I see him sit without a murmur, without a symptom of discontent, in a state which I deem wretched; when I see him contented and satisfied with the simplest support which nature can bestow, evidently enjoying a situation which has to me not a charm, which has not a luxury in it; I learn how difficult it is to judge, and how much I ought to consider before pronouncing on any state; I feel how true it is, that there is no condition without comforts; that indeed blessings have been distributed to all mankind. I see that man indeed wants but little, that in seeking for much he only increases his wants; and I see that it is not in the splendid mansion, or at the richly-covered table, or even in the very enlightened mind itself, that happiness is, alone, to be sought for. And seeing that it is so difficult to judge of a state, I learn that I should neither envy those that may seem far above me, or contemn those that may seem as far beneath me; but that I should look with complacency on all, and endeavour to think my lot, whatever it may be, as probably the best.

This is the truth, and it is good that it is so; every heart turns to its own home, and sees there the most enjoyment: every one esteems his own country the best, and the comforts he has in it the greatest. Although there may be some that we think wretched enough, luckily, they do not think so themselves; and though we may suppose that by changing their state we would add to their happiness, the probability is, we would

H

take from it. The frozen-up Laplander, and the sun-burned Indian are happy in their own climes; inhospitable as they are, they are happier in them, by far, than they could be in any other; and just from the cause, that there their affections have grown and fixed: and upon these all their happiness depend. Our affections get rivetted upon what infancy presents to us, and whatever it may be it seems lovely, and sorrow must be the consequence of taking us from it. Though it be but a humble cottage in a barren secluded situation, it will have its thousand charms, and there is no scenery, however rich, no mansion, however splendid, the heart will dwell upon with so much fondness.

And the Britainer, however long he may wander, however far he may travel, never will he find a country, which, give it him in his acceptance to choose for life, he would prefer to his own. If curiosity, or necessity, or perhaps discontent, has induced him to visit countries most famed for their beauties and luxuries, yet will he say when he has seen and tried them, that they have not the beauty of his native isle; that they possess not comforts half so true. Wherever he goes his thoughts will continually turn to the land where he first drew his breath,—to the land where he received his earliest, his happiest, and ever-during impressions. Place him in a situation the most alluring; give him all the comforts a foreign country can give, and still there will be a sighing after the country that gave him birth. Let him only feel that he is for ever removed from it, and there is a wonderful sadness produced. However long he may be absent, that thought is ever the greatest and the most cheering; that he will yet live and die in his native land. And thus are we none the worse for wandering; for the farther we roam, the better will we find out, and the more will we be convinced, what our home really is. And still more will we be bettered by it, if a discontented spirit has sent us a roaming: for the likelihood is, that we will return with the spirit cured, and confess that our notions were mistaken, that we went to seek that in another country which existed most in our own.

How touchingly does Campbell, in his 'Soldier's Dream,' advert to this native preference and fondness.

Methought from the battle-fields' dreadful array,
Far, far I had roam'd on a desolate track;
'Twas autumn,—and sunshine arose on the way
To the home of my fathers, that welcom'd me back.

I flew to the pleasant fields, travers'd so oft
In life's morning march when my bosom was young;
I heard my own mountain-goats bleating aloft,
And knew the sweet strain that the corn-reapers sung.

Then pledg'd we the wine-cup, and fondly I swore,
From my home and my weeping friends never to part;
My little ones kiss'd me a thousand times o'er,
And my wife sobb'd aloud in her fulness of heart,

'Stay, stay with us,—rest, thou art weary and worn!'
And fain was the war-broken soldier to stay.

In a very few weeks after we arrived at Calcutta, I had parted with almost every passenger we brought out. All, I believe, except two, had left Calcutta, and almost all had gone different ways. It may be, they and I are never to meet again. I may have parted with them then for the last time, and the happiness their society once gave me it may give no more. But whether we are destined to meet again or not, they shall ever live in my remembrance. Wherever they go, my warmest affection, and my sincerest wishes go with them; and until the last hour of my life, shall I look back with comfort, with a pleasing regret, upon the time we spent together.

PART II.

INSTRUCTIONS FOR THE PRESERVATION OF HEALTH IN INDIAN CLIMATES.

> Some (as thou saw'st) by violent stroke shall die;
> By fire, flood, famine: by intemperance more
> In meats and drinks, which on the earth shall bring
> Diseases dire.
>
> MILTON.

It is hardly necessary to give any directions regarding what is to be done for the preservation of health during the voyage from England to India. For then the health is generally very good; it is often good even in those who have on shore laboured under disease, whether disease coming of itself, or disease brought on by improper conduct; and this for evident reasons. Now there is a pure, and consequently salubrious, atmosphere, breathed; and what is perhaps as much to be wished for and admired, there is no excess committed. The invalid now comes to shew unusual symptoms of strength; he who on shore never shewed any inclination for food, now comes to eat with an appetite; for the pure air he breathes, and the other causes which have been brought to bear on his system, have wrought a change, and he is unusually invigorated and improved. The irregular liver is now altogether removed from the field of irregularity; he cannot if he would, unless he transgresses all the bounds of good society, enter now into the excesses which he may formerly have indulged in; and the good which such a restraint has on him is soon very palpably perceived. In the generality of ships sailing from England to India, the greatest regularity is observed; no impropriety is allowed; no debauching or late seats are ever countenanced; he that would lead a life at all at variance with what propriety points out, must lead it apart from these he is associated with; and to this must be attributed, in no inconsiderable degree, the general good health of those who go an Indian voyage.

The only direction which I find myself called upon to give is, to pay attention to the state of the bowels. It often happens that, from change of diet, want of the usual exercise, and other causes, people when they first go to sea get the bowels considerably constipated; and this it is highly necessary to attend to and obviate; for the rest of the system can never be altogether right when these are out of order. A pill, composed of five or six grains of calomel, with three or four of the extract of colocynth or jalap, taken at night, and a tea-spoonful or two of Epsom salts, dissolved in a tumbler of water, or the salts with senna, if the person has formerly been in the habit of taking them so, or a table-spoonful of castor oil, or a tea-spoonful of the compound powder of jalap, taken in the morning, will be the proper medicine where an active purgative is required. But when only something gentle is wanted, merely to keep the bowels free, the pill itself, every evening, or every second or third evening as required, or the tea-spoonful or two of salts every morning or second morning, or, occasionally, a little rhubarb and magnesia, or cream of tartar and sulphur, or whatever the person may have formerly used, and found best to suit him, will be quite sufficient. But the truth is, if regularity be observed, and all the exercise taken which may be taken, there will be but little need for medicine of any kind. All the functions will go on well enough, and when they do so there can be no necessity for interfering with them. With some people, fresh prunes, figs, and things of that sort, answer all the purposes of medicine; and when that is the case it is fortunate, for not only is the medicine then taken without disgust, but it acts as a nutriment as well as a medicine; a thing much to be wished for in those who have not much appetite, especially in the earlier part of the voyage.

But it is a fault in many who go to sea, that they do not even take that exercise which they might take. It is true that they cannot take the same laborious exercise which they may have been accustomed to on shore; they cannot ride, or walk a stretch of miles in the same direction; but they can have, I venture to say, in general, almost as much motion as health requires. They have the range of the ship; and this for him that is at all inclined to be active, is latitude enough. But the fact is, many when they get on board ship, get a little lazy, and are inclined more to lounge than to keep in motion—putting the blame upon the place which attaches to themselves. But he that has a proper regard for health of body, as well as buoyancy of mind, will not neglect this very necessary duty. In the morning and evening, and occasionally in the day when he feels inclined, let him take his stroll on the quarter-deck; and not only will he find that thus he passes lightsomely away half an hour or an hour of the long day, but at the same time he is invigorated, and when he comes to rest, he can bend both body and mind to employment with two-fold ardour.

As I am speaking on the state of health experienced in a sea voyage, and as we had a case in the Lonach a good deal in point, I cannot help throwing in an observation or two, in passing, regarding the good effects sometimes accruing from the removal of consumptive patients into a saline atmosphere, and from a colder into a warmer climate. It is now very generally, I believe, admitted by physicians, that when consumption is actually threatened, perhaps the best chance which the patient has of recovery, is by a removal to a more genial climate; and whether it be the mere change of temperature, or the peculiarities of the air, or the change of scene, or all put together, which work the good effect, it matters not much to inquire; it is enough that we know a sea voyage and a tropical temperature *are* often productive of good effects in the disease —this should be enough to induce those who stand in need of the remedies to give them a trial. I do not at all advise a removal to *India* for a consumptive European patient. The voyage is in the first place too long; the changes of temperature experienced in it are too many; and therefore it is only those that have something or other to call them there—that have something to induce them to choose that place in particular—that I should ever advise to such a voyage. I merely speak of the matter because it is in some degree connected with my subject, and because I should wish, if indeed there be any means of arresting the march of such a melancholy and ravaging malady, to lend my feeble aid in making them public.

When I saw Mr. C—, who went out passenger with us in the Lonach, at Gravesend, the night before we sailed, I thought I had reason to fear, from his appearance, that he might not see the end of the voyage. He had so many of the symptoms of threatened consumption—the severe cough—emaciation—quick pulse—hectic countenance, *&c.*, that some would have been inclined to say, his disease had indeed gone too far to be checked by any means whatever. But scarcely had we got fairly out to sea, when there was an evident change for the better. The cough diminished much, and the appearance altogether became healthier. As we got into the warmer climate, the improvement gradually went on; and by the time we had been a month at sea, there was not a symptom of his complaint remaining. I dreaded that a change of temperature again might be apt to throw him back—that the cold we might experience in the high southern latitudes of the Cape of Good Hope, would be in some degree injurious to him;—but I cannot say that even there, there was the least return of the disease. We landed him in India without a complaint; and when I reflected on, and contrasted his general appearance then, with the state he was in when I first saw him, I could not help being much impressed with the excellent effects which seemed to have accrued from the voyage. It is but right to add, that Mr. C— was an exceedingly regular young man: and nothing that could at all tend to promote recovery, was neglected on his part.

I know that a great bar and objection exists to the adoption of the remedy now spoken of, even in those who can most easily have recourse to it—I mean the reluctance to journey so far from home—to sacrifice, and while in a sickly condition too, when their aid is most needed, the society of perhaps every single friend we possess, —and the fear, that even after all the sacrifice no benefit may follow: and I acknowledge the objections are not undeserving of attention. It is indeed hard to part, at such a time, with almost all that gives us comfort; and it is true, that not a few have gone never to return—to die strangers in a foreign land: for it is but too well known, that when the disease has proceeded a certain length, no remedy which we yet know of, no plan of treatment which we have yet tried, will have any effect in stopping its fatal course. To urge patients, with the complaint thus far advanced, away from home to a distant country, is not only foolish, but cruel. We must, to have any good chance of overcoming it, catch the complaint in its very beginning; we must, it is true, see it ere it has made any progress, and take the proper steps then. If we are to advise a removal at all, the sooner, undoubtedly, we advise it, the better. But surely, so long as the hope is considerable, that the disease is still curable, the patient should bring himself to the terms even as they are. When he is told, that there is nothing else at all likely to be of use; that though the hope of recovery from a removal is not altogether sure, there is much less hope if he stays where he is; surely if he has it in his power to have recourse to the remedy, considerable though the sacrifices may be which he has to make, he ought to make them, and give it a trial. The physician, of course, is to be the judge, and to pronounce on the case. He can, in general, tell pretty correctly what the probabilities are—what the chance of the case is—and, of course, just by the advice which he gives, is the patient to be guided.

The different West-India islands, the island of Madeira, the Canaries, *&c.*, have all occasionally been chosen for the residence of the invalid. Circumstances generally direct to some particular one of these, and as generally may the patient be allowed to do in this as suits him best; and when it happens that circumstances will not admit of a permanent residence in any part, still there may be much benefit derived from merely the voyage. The cure is often, indeed, nearly complete sometimes before the destined port is reached; and therefore, though merely the voyage can be taken, let it certainly be tried. A voyage to any of the above-mentioned places, or a cruise for a little while on the southern coasts of Spain and Italy, if that is convenient,—or a still shorter journey, even a trip from one British port to another, may be attended with much advantage; for, as I have said before, it is not perhaps the removal from the cold into the warm region altogether which does the good. We often see the evident change some time before we get into the warm climate, and therefore, if something depends on any motion of the vessel, as indeed it seems, or on the peculiarities of the air, or on the new scene the patient is introduced to, even the shortest of voyages may not be without its good effects.

I would just add, that it is not only in consumption that the remedy now spoken of, and in particular the shorter voyages, may be useful.

I

Some of those ailments peculiar to the female, which, indeed, end not unfrequently in consumption. The complaints denominated nervous or hypochrondriacal; the ailments so commonly attributed to the stomach, which afflict such a multitude in society, might all, I think, be benefited by the plan. Not a few have found out, and just by accident, perhaps, that a proper attack of sea-sickness has done away with all the ailments they were formerly liable to. There seemed to be a total change wrought in the system by it; and, therefore, do I recommend all who stand in need of the remedy to give it if they can a trial.

But to return to the main subject. It is when we have landed on the Indian shore, that we become especially liable to disease, and it becomes especially our duty to guard against it. Then, however healthy we may have been in our native land, and however vigorous during the voyage, it is necessary to remember that we stand on ticklish ground; there is now, undoubtedly, an unusual liability to attacks of disease; and as these attacks are both sudden and acute, cutting off without discrimination both the strong and the weak, it is well to take care that no conduct on our part shall increase this liability. There are some who deny, that the European constitution is at all more obnoxious to disease in a foreign climate than in its native situation. They say, "let us go where we will, and let us live how we will, it is all the same, we will not die till our time comes;" and, as a proof of it, they adduce the example of those few who have so travelled and lived, and have shewn all the marks of health at the end of a long pilgrimage. But this is mere foolishness; the proof adduced is an exceedingly bad one, for it is the few out of the multitude. We have seen quite enough to convince us, that a little care will sometimes save, when the opposite plan, a little imprudence, will destroy. But there are some reckless of their constitutions, and every thing else, in all situations, both at home and abroad. I feel just as much assured, that the European transplanted to an Indian climate, becomes peculiarly liable to disease, as I feel assured, that this earth is a planet revolving round the sun; and it is this assurance which makes me press attention to the matter so earnestly. Far be it from me to infuse unnecessary fear into the bosom of the stranger landing in a foreign country. He needs, at such a time, I know well, all the support that can be given him on many accounts; and I would not have the thought of the danger he is exposed to from disease press on his spirits for a moment. I would have him firm as a rock upon that head; but I would have him also just keep in mind that such a tendency to disease does exist, that he is now not just so sure of health as he may formerly have been; that he must not now indulge just so much as he may formerly have done, even with impunity; and he will be none the worse for keeping all this in mind.

And I am glad to say, that by the majority of Europeans in India, the right view of the subject is taken, and the proper plan followed. Their manner of living is exceedingly regular;

and I may well say, if the new-comer is guilty of any impropriety, he has himself alone to blame for it, for he will not be urged to it by those he is associated with. There are, it is true, a few to be met with, who will carouse, and debauch, and sit late, even in India. They are, however, but few, and the wise avoid them. In all good society, though the table be, indeed, richly covered, few of its delicacies are indulged in. Even the seasoned European, he that has by residence become in some degree naturalized to the climate, will not exceed; and the new-comer will but seldom be urged to go even so far as him. At an early hour the table is left, and at an early hour they retire to rest. In the morning again they are timely up, taking the exercise which is, as it were, to invigorate them for the day; and just by this regular system of living multitudes enjoy a perfect and long-continued state of health. The only fault I would find is, that the table is covered rather richly. There is rather too great a profusion, rather too many delicacies, many more than is required; and the stranger may thus be induced to exceed unwarily. I must own also, that he is occasionally pressed a little too much to the profusion. Fashion, no doubt, requires the superabundance ; it is just what she expects and must have. But it is a pity fashion is just so much observed ; her dictates are in some points extremely injurious ; and it would be well if sometimes she was more firmly stood against. It would surely be better, if less to entice and goad on the appetite was placed before us ; and at all events, there should be no urging to more than we feel satisfied with. But the prudent will always withstand both inclination and invitation to indulge in more than enough.

I need not be entering too minutely into directions for the regulation of the diet ; for common sense, I think, is quite enough to teach this ; and when I generally recommend moderation, or indeed a little abstemiousness rather, especially at first, I think I give almost all the lessons which the prudent require. I may just, however, give a short view of the Indian customs in that respect. The breakfast is almost invariably at eight; and it is in general a good meal. The early rising there gives a good appetite in the morning; and he that is in good health, he that has not been living improperly the night before, is always well prepared for the morning repast. With the addition of rice, which is always a principal article of Indian food, the breakfast is just the European one,—tea, coffee, *&c.*, the articles are both wholesome and substantial, and well enough suited to the European constitution. At one or two o'clock, with the greater number, there is what they term a *tiffin*, which answers to our mid-day light meal, or *lunch*; it is generally just something light, a little curried rice, or some bread and fruit, *&c.*, with a glass or two of wine. Then at seven in the evening comes the dinner, which, excepting perhaps a little more superfluity in it, is much the same as an English dinner; and this is the last meal. Suppers may occasionally be given; but generally speaking, we may say the dinner is the supper; for by the time it is concluded, the hour of retiring to rest (which is ten) is close at hand. Some, whom it suits, prefer having the dinner at three or four, with no tiffin, but tea or coffee again in the evening; which is thought by many to be much the best plan, and on this account,—that then we do not go to bed till a long time after the full meal, and therefore are likely to have sounder rest. And this may be true in some degree. It may indeed be as well, when the system of dining at the early hour suits best, to adopt it. But the truth is, we should never take what will incommode us; we should never either eat or drink what we need be afraid to go to sleep with. He that takes no more than appetite requires, may lie down safely enough, even the instant he has feasted; and therefore he who is a prudent man need not be afraid of the seven o'clock dinner.

It is a question not unfrequently adverted to; whether it be best, to drink at one meal only one kind of liquor, or to partake of a variety. There are some who say—Keep only to the one thing, whatever it is, and you need not be afraid of going a considerable length; while there are others who contend, that you may just as safely take all and every thing that is presented. But the truth is, both parties are very wrong, there is, undoubtedly, much error of both sides; we know well that the man in good health absolutely requires no kind of liquor whatever—the system, naturally acting, stands in need of no such stimulus to keep it a-going. But as society has introduced the custom of taking certain beverages at meals, we must perhaps, while we continue to live in society, be content to depart a little from nature's strict rule, and concede in so far to the custom; yet, if we indulge beyond a certain length, even in one thing, we will certainly be injured by it. If we take up the idea, that so long as we keep by the one liquor we may drink deeply without fear, we lay down for ourselves a most erroneous and ruinous system; and on the other hand, when we indulge in variety, we are exposed to this great source of error, that we are apt even unawares to indulge too much. This seems to me to be the great bane in the variety of our drinks, as well as our meats—that too much is presented to the palate—too many things to entice are offered, and thus in truth, we eat and drink by far too liberally. I am satisfied that the various articles of a fashionable dinner, though not perhaps more liberally indulged in than we would indulge in those of a plainer one, will not be just so wholesome to the system as simpler fare; yet, if we only indulge in them till nature has said we have enough, I do not think we need at all fear the consequences. I am satisfied that he who drinks only Madeira, or claret, or beer (which seems now to have become the most favourite beverage of India) may not so soon feel injurious effects as he who takes all of them—the very mixture, it is true, of the different liquors may be pernicious; but, as I have said, I think it is more from the circumstance of the general drinker's being apt to go too far, that danger is to be apprehended; if the single drinker makes also free, why, he will suffer in time just about as much as the other. I certainly recommend strongly the adherence to one liquor; for it must be owned a multiplicity even in very small quantities, must be particularly injurious; and there is indeed no necessity for more; one serves all the purposes of society. But if we are ever induced to venture upon more, let us take care that we venture but a certain length. He perhaps that takes none does best; he that takes but little is surely next to him in prudence.

There is another very erroneous idea got amongst Europeans in India. There is not one in twenty of them that will drink pure water; they must either have a little brandy, or wine, or spirit of some kind or other in it; and it is enough almost to frighten some, to see a glass of pure water taken off at a draught. Now, the water seemed to me to be very good indeed, and therefore I was anxious to ascertain what was the cause for all this antipathy. But the majority could give no reason, except that water by itself was generally allowed to be injurious; and a few explained it to be the saltpetre with which it is impregnated. I do not know, indeed, whether there really is saltpetre in combination with it or not; but saltpetre, in that quantity, I should think not such a mighty injurious article. And does the admixture of brandy take out the saltpetre, then? is this not just adding one injurious article to another? Oh! the brandy corrects it, ay—at all events it is not a disagreeable corrector, and whether it answers the purpose or not, *it is taken without much disgust.* I am convinced no one need be afraid to drink pure cold water when inclined, unless previously heated by exercise, or some other cause to forbid it; and I decidedly enter my protest against that habit of drinking so continually brandy and water, or *brandy pany* as they call it. In a climate such as India, we naturally crave for a considerable quantity of fluid, and if we mix spirits always with what we drink, we must undoubtedly take what will sooner or later be hurtful; and let it be recollected, as has already been hinted at, that it is a thing which grows, that though we take it only from example, to be like others, at first, we come in time to take it just because we like it. If we must mix, let the quantity of spirit be indeed small; we can pledge our friend in a glass gently tinged, as heartily as we can do in a *stiffener*. Many, I am satisfied, have ruined excellent constitutions just by this habit of drinking,—and

> " What then avails, that with exhaustless store
> Obsequious luxury loads thy glittering shrine;
> What then avails, that prostrate slaves adore,
> Or fame proclaims thee matchless and divine!
>
> Can gaiety the vanished years restore,
> Or on the withering limbs fresh beauty shed!
>
> In health how fair, how ghastly in decay,
> Man's lofty form!"

Next to a proper attention to diet, a proper attention to exercise is required. There is nothing, perhaps in any climate, so likely to keep up the proper balance of the system, and keep all its parts in healthy action, as a due quantity of motion; nothing so likely to let the system fall, as an undue inactivity, and therefore to take all the exercise we can, should ever be a chief point with us in India. Of course, there it is impossible to be much in motion during the day; the heat is then certainly too oppressive. But in the mornings and evenings, and especially in the cooler months, exercise can very well be taken; and to a certain extent it very generally is taken. Almost every one keeps a horse, or a vehicle of some kind or other. There are few indeed, who have not the means of doing this; and from day-break till the sun is well above the horizon, and again in the evening for an hour after he has gone down, all are seen out taking the accustomed ride. This, to be sure, is so far well; but I do not think this is enough of exercise, or as much as might be taken. It is the fashion to ride, it is not much the fashion to walk, or walking for the same length of time would have better effects. There can be no doubt but walking is by far the most natural kind of exercise; and surely for him at least, who has perhaps for the first fifteen or twenty years of his life been in the habit of walking some miles every day, a little of it is requisite even in the warm climate, and ought perhaps to be preferred to any other mode of taking exercise. There must be a change produced in the system very much for the worse, if from such a state of action we enter into one of almost perfect inaction; if we do as many do, scarcely ever put a foot to the ground. So that the European who is anxious to preserve health, when he becomes a sojourner in India, though he may have it in his power to ride when he chooses, will also occasionally make use of his own limbs; he will not forget altogether, for all fashion may set before him, his earlier and more natural habit. We can walk, too, at home in our house—under our *virandah*, often in seasons when the weather does not admit of riding out.

There is, however, one kind of exercise which I do not much recommend. I mean sporting. In such a country as India, we would hardly expect this recreation to be in use. But it is true that the gun and dog are there in very considerable demand. There are not a few to be found, who will not only engage in the sport during the cooler parts of the day, and in the cooler seasons, but who will engage in it at all times, and in all seasons. It is a general belief, and I dare say a belief founded on good observation, that exposure to the strong rays of the sun during the day, is apt to be productive of very serious disease. It is, I believe, too true, that many have untimely come to their end by such an exposure. But whether it be true to the full extent or not, even allowing this dread of the sun to be a little more than is necessary; still, it is well, when we can, to keep on the safe side, and not wantonly throw ourselves in harm's way. But with the sportsman all fear seems banished; although at other times, he would be afraid even to put his head out of his *palanquin*, or carriage, if the sun was appearing, yet when he enters on his favourite amusement, he minds sun and danger no more; but travels heedlessly on through jungle, and marshes, and paddy fields, exposed to damps and exhalations of every description, and thus undoubtedly may perish. When an hour or two's shooting, morning and evening, can be conveniently taken, it is all very well; but no danger should be encountered for it. And I would also add, that when we do sport, we should not, as some do, have recourse to the *brandy pany* to quench thirst; for this perhaps kills about as many as the sun does.

Neither does dancing seem an exercise well suited to India. In England, and climates like it, I do believe dancing, in a moderate degree, to be a healthy amusement. But in countries within the tropics, I must deem it prejudicial even in any degree. Yet, in India it is about as common, I think, as in Europe. I have seen in Calcutta, and not in the coldest season either, male and female a full hour in the dance; and no doubt such a system may sometimes be followed for a considerable time without much appearance of injury. But it cannot be without bad effects; the chance is that sooner or later its ill consequences will come to be felt; and therefore the prudent will either altogether avoid it, or at any rate indulge very cautiously.

The clothing worn in these climates is of course

of a very light kind. In the warm seasons, a jacket, made of white jane or something resembling it, with vest and trowsers of the same kind of stuff, is the general wear. In the cold seasons, and particularly in the more northern parts of India, clothing considerably warmer is required. At Calcutta even, which is not just out of the tropic, in the months of November, December, January, and February, habiliments not much differing from our English ones, can very well be borne, at any rate in the mornings and evenings. Cotton shirts are always preferred to linen. But on the head of clothing it is unnecessary to say much, as the tropical visiter is very soon taught by his feelings what suits best in this respect. Should he arrive in the hot season, if he be still fond of his European dress, as some are, and think to wear it, he will very soon be glad to follow the example of others, and exchange it for the Indian one. And, on the contrary, should he arrive in the cold season, and especially if he is despatched immediately off to some pretty northern situation, though at first he may mount the light Indian garb, he will soon come to find out that some of his European articles will suit him fully better; and likewise, he will find that a blanket around him at night is by no means a disagreeable thing.

It is a question very frequently asked, whether in tropical countries it is requisite to wear flannel; and the answer I would give to it is, that those who have never worn it before, and who are in perfect health, perhaps need not put on flannel; but at all events, if it has been worn before, and especially if it has been put on for any ailment, it should never on going to a warm country be left off. It is supposed by many, that in such a temperature it must be dreadfully cumbersome and disagreeable. But this is not the case; it is only on the body the flannel is required; and when the shirt is thus made without sleeves, and of the lightest kind of flannel, it adds exceedingly little to the weight. It may annoy rather for a day or two at first, but afterwards it is never felt; and as it undoubtedly is of much service, keeping off affections of the bowels and numerous other ailments, even if it did give a little uneasiness it should be borne. As a substitute for flannel, many are in the habit of wearing under the common shirt, a light calico vestment; which is also a very serviceable article.

Another great mean for the preservation of health in India, is, the regular use of the cold bath. Every house has its bathing place attached to it; and the manner of using the bath is, either to stand up, or sit on a chair, while another pours four, six, or eight potfuls of water over the head and body. It is a most grateful thing, after a warm restless night, thus to get fairly soused and cooled; whilst at the same time it invigorates the system much, and serves to fortify it strongly against disease. I am satisfied it preserves many in health, who would otherwise have but very indifferent constitutions. Some take the bath two or three times a day; but I think once is enough; let it be taken regularly in the morning.

As to medicine, if attention be given to the

proper manner of living, not much of it will be required, even in India. There are some who must be continually taking their calomel, *&c.*; and, though I would be far from checking in any a proper attention to the state of the system, still, as I think, that the habit of calomel dosing at one's own hand, when once begun, is sometimes carried too far, and used too indiscriminately, I must rather advise that we meddle with ourselves as little as possible. Certainly, when the system is felt to be out of order—when disease seems to be threatened—have immediate recourse to the proper means. But then I would recommend that the patient do not trust to himself, but should have the advice of those who know best how to treat him. Certainly, let not a moment elapse ere we interfere, when interference is really demanded. But, as I have already said, if we guide ourselves right, we will the seldomer need the doctor; if regularity was just a little more observed by some, there would be less need for calomel.

And, though I have mentioned it last, perhaps

not the least mean for the preservation of health in a foreign country, is, a firm and contented mind. There has been quite enough seen to prove, that any sudden and great alarm, any unusual anxiety about ourselves, a great dread of disease, for instance, actually predisposes to disease. It cannot, indeed, be said, that any degree of fear will absolutely and certainly bring on disease; but it has been so repeatedly seen, during the rage of the epidemic, that those who stood fearless and undaunted amid all its ravages, were much less liable to come under its malignant influence, than those who were every moment pondering upon it and expecting its fatal attack, that with good reason we may recommend a steady mind as no inconsiderable preventive of malady. It is easy enough to see how it happens; the body, weakened and paralyzed by fear, comes in an especial manner to be *susceptible to the impressions of disease.* It is well enough known, how much the body is influenced by, and sympathizes with, the mind; and therefore should it always be our endeavour, to keep the heart well up, whatever fatalities may be occurring around us. We are not, indeed, from any fool-hardiness, to brave or court disease; but we are as carefully to guard against an over-timidity. If it is to come, why, it is a pity to anticipate it; if it is not to come, as foolish is it to give ourselves unnecessary pain. And, at the same time, I would caution against that anxiety, or discontentedness, which we see so frequently attaching itself to the mind of the foreign sojourner, arising merely from his change of situation. He is not just satisfied with the place he is in; he has, perhaps, not found it just what he expected; there is a sighing after home—a continual looking to where he was—preventing him in a wonderful degree from being at all happy where he is; and though anxiety, or dissatisfaction, may not itself cause disease, still it may assist; it at all events disposes us to that languid inactive state which favours the coming of disease. And what good can it do? does it not only keep us continually unhappy? What can all our sighing avail? we are now in the foreign land, and here we must be, and we must just make the best of it. Surely it is better to be happy in some degree, though we may not be just to the degree we could wish. Therefore, let us endeavour never to be much depressed without some very good cause; let us bear up with a bold heart even against the burden that may be pressing on us. And to accomplish this rightly, our best plan will be, to train and occupy rightly both body and mind. Let proper employment, with proper recreation—exercise—study—a book—a cheerful friend—all in their turn, be attended to by the foreign sojourner.

"From labour health, from health contentment springs;
Contentment opes the source of every joy."

And

"True dignity is his, whose tranquil mind
Virtue has raised above the things below;
Who, every hope and fear to Heaven resign'd,
Shrinks not, tho' Fortune aim her deadliest blow."

"Thro' the perils of chance and the scowl of disdain,
May my front be unalter'd, my courage elate;
Yea, even the name I have worshipp'd in vain,
Shall awake not the sigh of remembrance again:
To bear, is to conquer our fate!"

To sum up the whole, and bring it into one short view—let temperance, in our meats as well as in our drinks, be always remembered; we are on no account, whatever may be the variety or quality of our food, to exceed the proper quantity, or overstep the point where nature makes her stand. Let us take all the exercise we can, stopping short of fatigue; but not that exercise which will either heat too much, or expose us to the strong sun, or to damps, dews, *&c.* Let us clothe just as the season points out;—while at the same time, we take our regular bath in the morning—attending also, but not with too much anxiety, to the state of the bowels, *&c.*, and keeping a stout heart amid all calamity—never allowing the spirit to sink from fear, or to be inactive and unhappy from discontent.

I have thus endeavoured to bring forward as clearly, and in as few words as possible, some of the lessons most necessary to be known by the stranger taking up his residence in India. I have not entered so minutely into the subject as I might have done; I have not remarked upon some little points which might have been remarked on, because I think it needless. He that is anxious to attend to the rules laid down, will not be likely to err far in any thing else; a little reflection must in almost all things point out what is proper, and he that is really inclined to act well cannot be much at a loss to decide on his line of conduct. And I am fully satisfied, as I have asserted in the outset, that an attendance to such rules as these may be attended with no small benefit. It is too well known, that disease will attack all, and cut off all—the temperate as well as the free; but it is as well known, that the temperate have the best chance either of escaping altogether, or of recovering when they are attacked. It would be saying what was false to assert, that the climate itself is not enough to break up the constitution—that long residence, even under the most temperate living, is not often the source of disease; but it may with good reason be asserted, that the climate is sometimes charged with what it does not deserve —that many of those who return to their native land, half dead with disease, have their own conduct, and not the long residence, to blame for it.

PART III.

HINTS TO SURGEONS AND OWNERS OF PRIVATE TRADING SHIPS.

All the world's a stage,
And all the men and women merely players;
They have their exits and their entrances,
And one man in his time *must* play many parts.

SHAKSPEARE

In ships of war, and in Indiamen, there are certain places allotted for the sick, and certain persons appointed to attend upon them. Immediately a patient applies, however trivial his complaint may be, if the surgeon deems it necessary, he is put upon the sick list, and kept there till the cure is completed. And although a considerable number be taken ill at the same time, they can all, without any difficulty, be taken from duty and laid up. Then when the surgeon has his patients laid up, he has only to visit and prescribe for them, and whatever he orders to be done, he knows will be done. Whatever he thinks it necessary to advise, either in the way of medicine, or food, or any thing else, he knows, without any farther trouble to him, will be strictly attended to by those whose business it is to wait upon the sick. And thus he has both success and pleasure in his practice. Just as in an hospital on shore, he has his patients completely to himself; he has every thing done for them just as he would wish it; there is nothing to thwart him, no difficulty standing in his way.

But in the small merchant-ship it is very different. In it there is but little room, and every place that can at all be spared is occupied. There is no place for the sick, but the common place for all; neither is there any one to attend upon the patient during his sickness. There are no more men than just enough to do the duty of the ship; consequently it is difficult to spare, particularly at certain times, even one of them.

And thus the surgeon is by no means agreeably situated. He has frequently to allow men to remain at duty when he thinks it necessary they should be taken from it; and thus combating the complaint under disadvantageous circumstances, he does not get things done, and his patients do not get on so well as he would wish. And when circumstances absolutely demand that the patient be sent to bed, even then he does not get free of his difficulties. There being no one to give the patient any help, and the patient perhaps not very able to help himself, the surgeon himself must do every thing, or scarcely any thing will be done at all. And frequently when he has done all he can, he will not have done all he could wish. There are annoyances in his way which he cannot overcome; and neither he nor his patient get fair play.

One instance of difficulty in the practice of the small-trading ship is, when ulcers occur upon the legs or feet, a complaint not uncommon among seamen. So long as the man remains at his duty, and exposed to all weathers and usage, it is sometimes no very easy matter to get him cured, it is hardly possible to have proper recourse to all the necessary measures. But as the man is probably otherwise in good health, and as able to do his duty as ever, he cannot be allowed to lay past for such a small thing; and the cure must either be accomplished tediously while he remains at work, or perhaps after all he is obliged to be laid up.

Another is, in cases where we wish to give mercury. For all the arguments of the anti-mercurialists, most men yet, when they get a patient with what they conceive to be a truly syphilitic affection, never think the patient altogether safe, although the more prominent symptoms may disappear, unless they have brought his system under the influence of mercury, and kept it so for a time. Yet as the seaman in the private-trading ship is so absolutely needed, although he really has syphilis, as he is nevertheless able enough for duty, he will not be excused for such a considerable time; and we are under the necessity of administering the mineral exposed as he is, or perhaps have to prefer going in, in some degree with the plan of the anti-mercurialists. In warm latitudes, and in fine settled weather, we can venture without much fear, to give the medicine even though the patient be at duty.

In these ships there is no *commode* for the convenience of the sick; and this is a great want. It must undoubtedly be often productive of very injurious effects for a man labouring under disease to rise from his warm bed, probably frequently in the night, and probably when it rains hard and a piercing wind blows, and go to the head of the ship, where the common commode is. Yet this he must do, for there is no place else. And from this cause are we often obliged, when the weather is bad, to put off for a time giving medicine which we are inclined to give. But the case sometimes demands immediate attention; and when the disease is dysentery, the circumstance is exceedingly unfortunate. In such cases, in cases where we suppose there is danger, disagreeable as it is for the healthy,

L

the patient must not be allowed to leave the birth.

Emetics are now not so frequently prescribed in diseases of warm climates as they have been, but still cases occur now and then where we think they are called for, and when these cases happen the surgeon must in general be content to take a little trouble. As no one else thinks it their business to assist the sick man, and as the sick man in many cases must have assistance, the surgeon cannot refuse to become the assistant. He must push about and get whatever is required; he must indeed, if he does what is right, superintend the whole operation,

There is also some difficulty occasionally experienced when a patient is recovering from disease; there is a difficulty in getting for him certain things in the shape of food or drink which we think he requires. It is well known that a person just getting the better of an acute complaint, requires a little nursing. The stomach will receive and relish, when weak, many kinds of light pleasant nourishment, though it loathes the common fare of sailors ; and much can often be done for the patient, the cure can sometimes be considerably expedited by a little attention to diet. Yet no provision of this kind is made. There is nothing whatever laid in for the sick. And though in almost every ship, such things when they are really needed, can easily be got, still it is a little inconvenient not to have them. And even though they can be got, there is often from the want of proper assistance and regulation, some difficulty experienced in the preparation of them.

He, then, who goes surgeon of a private-trading ship must not expect to be altogether comfortable; he must not expect a practice so smooth as he may have been accustomed to on shore. I do not say he will have, by any means, a laborious situation, for his sick are never very numerous; but he will have rather a troublesome one. He must lay his account with meeting a good many annoyances, and annoyances too, which, with all the trouble he can take, he will not be able to get completely remedied. But

L 2

much certainly depends upon management; and though in many respects he is placed a little awkwardly, still, if he goes rightly about things, he will manage to lighten many of his grievances, and get on, upon the whole, tolerably well. Though he may find the situation a little irksome at first, as he gets better acquainted with it, he will get better pleased with it; the practice, which appeared at first the most disagreeable, he comes in a little time not to think so hardly of. If he looks about, he will generally find some one among the men possessing rather a more kindly disposition than the others, who will be ready to lend him a helping hand in his sick charge. If he manages well, he need not have any great difficulty in getting whatever things he may want in the shape of nourishment for the sick prepared; and from the liberality of almost every commander, he need not want when he requires it, either wine, or fresh food, or any thing else the ship can afford. However, as I have already hinted, he must take the most part of the labour upon himself. He must make up his own medicine; he must carry it himself to the patient, and he must see it taken; for Jack, even when ill, would rather dispense with the doctor's *drams*, and unless the doctor takes the trouble to see them fairly down the throat, the chance is, they never will go that road. He must see that the drinks, *&c.*, which his patient requires are really made and given; and if his patient be very ill, he must not think it too much even to attend occasionally and help him to them. In short, if he wants to have any credit by his practice, he must be apothecary and surgeon and nurse altogether; he must, in a word, superintend all and do all.

But surely it is incumbent on the owners of ships to send rather more hands than just enough to do the duty. Almost always one or two out of the number are unable for duty, and sometimes five, six, or more; and when this is the case, the labour comes very hard upon the remainder; indeed, they have sometimes considerable difficulty in getting the business carried on at all. The expense of a few supernumerary hands would be very insignificant; and as it would both enable the surgeon to take, without difficulty, a man from duty when he thought it requisite, and thus get him the sooner cured, and allow the business of the ship to be easily enough carried on, although two or three were absent: it is certainly a measure much to be wished—a measure which should never be neglected. How awkwardly must a commander be situated, if, as sometimes happens, he loses by death a few of his hands, when it is difficult or impossible for him to get others.

There should also be a boy shipped, whose especial business it should be to look after the sick. This need not be his only occupation; for frequently, particularly upon the passage, he would have but little to do in that way. He may be sailor or any thing else, as well as sick attendant; but he should be one that is not really needed for any other business—that can give, if it be required, the whole of his time to the sick. Such an attendant would be of immense use both to the patient and surgeon.

There should also be certain necessaries allowed for the consumption of the sick—a certain quantity of wine, beer, sago, and rice. It is not a great deal of these, I confess, that will be required. Stimulants and cordials are not dealt out now-a-days with such a liberal hand to sick folks as formerly; and seamen in general, when their complaint has been at once attacked and subdued, recover very well without much of their aid. Nevertheless there are cases where those least friendly to cordials will see it necessary to administer them. There will be cases where these things, either in the course or at the termination of the disease, are absolutely needed; and therefore they should surely be provided.

It could hardly be expected, that in small ships, where there is necessarily but little accomodation, and where all the accomodation is wanted, any separate place should be allowed for the reception of sick; nor indeed is such a place much required. I have, however, mentioned the great inconvenience arising from the sick having no commode close at hand; and this I think is a want which might surely be remedied. Such a convenience, for the sick alone, could easily be in the neighbourhood of their births, and the advantage of it would certainly much more than compensate for the expense and trouble.

The great moralizing poet says—

> Heaven speed the canvass, gallantly unfurl'd,
> To furnish and accomodate a world—
> To give the pole the produce of the sun,
> And knit th' unsocial nations into one:
> Let nothing adverse, nothing unforeseen,
> Impede the bark that ploughs the deep serene.

But, at the same time, he would—

> Teach *mercy* to ten thousand hearts that share
> The fears and hopes of the commercial care.

We wish to the merchant all the prosperity that can come to him; while, at the same time, we would wish, that he who "ploughs in his bark" should have all the comfort that can be given him.

A few of the surgeons of private trading ships, are in the habit of leaving the ship and living on shore, while she remains in harbour in India.

They visit their ship every day, but except at the time of the visit they are on shore. This however, although I adopted it myself, I do not think the proper way. The surgeon, having almost every thing that his patient requires, to do for him, he surely requires to be pretty often at hand; and if he is not at hand, which he cannot be if he lives principally on shore, it is very evident his patient cannot be properly attended to. Disease, too, attacking very suddenly in these climates, and committing great ravages in an exceedingly short time, is another good reason why the surgeon should never be much out of the way. Although he may visit his ship every day, or oftener, still this is not enough; he may not be there just at the time he is most required. And therefore should it be a rule of the surgeon's to live on board in port as well as at sea. As regards pecuniary matters, too, if he has any reason to be anxious on that head, he will find it his interest to prefer his ship to the shore.

And this naturally leads me to speak of a plan adopted by some owners of ships, not by any means a proper one—I mean the plan of engaging a surgeon going out to remain in the country in the service of the Company, to do the duty of the ship out, and take the chance of getting some Indian surgeon returning to England to do it home again. It must be obvious from what has been stated in the last paragraph, that this is indeed a very erroneous way of going to work. It is not upon the passage, either out or home, that the surgeon is much needed. So long as the ship is at sea there are generally very few complaints among the people; it is after she has reached her port, that complaints, and probably very severe ones, begin to appear; and, of course, it is just then that the medical man really is required; it is while she remains in harbour that she should have her surgeon, if she is to have him at all. I know that a medical man resident on shore may be engaged to attend while the ship remains in port; but from what I have already stated, it is evident, that it is quite impossible for him, let him be as attentive as he may, to give the attention that is required. He may visit and prescribe, but there is much more needed than this in the private trading ship. So that it should surely be a point with the owners of such ships, only to engage a surgeon who binds himself to perform the voyage.

Sailors, like most other folks, do not relish too much work, and it is alleged that, occasionally, when any extra labour is on hand, they sham sickness to escape it. This I have no doubt is sometimes done; but I am also of the opinion, that it is sometimes laid to their charge when they do not deserve it; and as they are fully more likely to play the trick in the small merchant ship, where, from not being too well manned, the duty is sometimes very hard, so, the man who gets sick then, if his complaint is not a pretty palpable one, and particularly if he is known to be a lazy kind of chap, is more apt to be suspected; and the doctor may occasionally get a hint to blister severely, or punish in some other way like this, those they suspect to be skulking. But this is a point where the surgeon requires to use much caution; and, if he does wisely, he will rather run the risk of being cheated, than needlessly give a man pain upon suspicions alone. He will generally be pretty well able to judge whether there be any thing wrong or not; yet, as he knows that a good deal may be wrong without him being able to see it very plainly, he will not, even in the case where he cannot discover much, form positive opinions against the man, or act rashly. If a man complains of head-ach, for instance, which he says is so severe as to incapacitate him from duty, although he cannot discover from the pulse, or any other index, that the head-ach really exists, yet knowing that for all that it may exist, and exist very severely too, we would act very unjustly and very inhumanly indeed, were we at once to brand the man for a skulker, and put him to unnecessary pain. In such cases by far the best way is, just to take the complaint for granted, and go on with the proper remedies; surely let us bleed, and blister, and purge, *&c.*, *&c.*; for this is just the plan likeliest to remove the head-ach if it indeed exists; and if it does not, why then he certainly deserves all that is done to him. But we are never to go farther than the symptoms he describes to us demand. We are never to give him the least reason to suppose that we suspect him; for if he finds this, he will perhaps, from mere obstinacy, determine to suffer all, and stand out against us. But we are just to treat him as he explains to us his complaints—and if he is a skulker, unless he is a very idle and a very strange fellow indeed, he will soon tire of our treatment, and be glad to exchange it for the ropes again.

It is a rule that when a seaman is put upon the sick list, his allowance of grog is stopped. It is supposed that then it is either improper for him, or that at all events he does not require it It is supposed too, that it prevents some from skulking who otherwise would. And this is all very true. But a sailor dislikes terribly to resign his grog. He looks upon it as one of his chief comforts; and for all we could lecture him to the contrary, he will continue to regard it as the best preservation of health he can get. And as there are some cases where we know all the spirit they are allowed can do very little hurt, if there is no particular circumstance to forbid it, the surgeon will perhaps do best not to adhere too rigidly to the rule. When the complaint is a chronic one, the patient an old man, and known to be a good man, I think he had as well just let the grog go on; he will not, I know, find his practice the worse for it. There is, however, some caution requisite in giving the indulgence.

Regarding practice itself, I shall not pretend to offer any thing. Mine of course has been very limited; and there are numerous and valuable sources whence the practitioner who enters upon the field of Indian practice, has it in his power to draw the knowledge which is to guide him till actual experience has taught him. I think, however, I may do what is not altogether useless, if I mention one or two of these sources. I may be excused if I just tell what authors I had, and from whom I thought I derived most instruction, when I entered upon the business. The only books I possessed, treating solely on the diseases of foreign climates, when I set sail in the Lonach were: "Mosely on Tropical Diseases;" "Lind on Climate;" "Clark on long Voyages;" and "Johnson on Tropical Influence." Were I to say that I studied the three first without advantage, I would say what is not true. They are the performances of men who practised in very considerable fields, of men of acknowledged ability, and undoubtedly they are well worthy the perusal of the tropical practitioner. In Mosely and Clark, in particular, there is much valuable matter; and though their practice be not exactly the one which practitioners now-a-days place most confidence in, still will we be able to glean something from it. In Mosely we find yellow fever, and in Clark the dysentery, treated by no means inactively; treated, though not just so boldly, not much different from the way we treat them now. It is not difficult, I think, to see that the method of cure which they adopted in these complaints, is the very ground-work of the more efficient one we adopt now; and later physicians, Johnson with the rest, are not a little indebted to them for the knowledge they now possess. Were we to shut out their practice altogether, they are worthy a perusal for their descriptions of disease.

I must however acknowledge that the three first authors were not absolutely necessary for me. Possessing Johnson's work, I could very well have done without the others. In it the most approved practice of the day is laid down, and exceedingly clearly laid down. It is in it that we find perhaps the best view of tropical disease that has yet been taken; that so intimate a connexion is traced between fever, and hepatitis, and dysentery, as almost to shew them to be only modifications of the same disease; that the same remedies modified a little, are applicable to all; and that they are not only applicable to them as occurring in the climate of India, but as occurring in all tropical climates, and very probably also in climates without the tropics. In it the surgeon is taught, that if he has the lancet, and calomel, and opium, and purgatives, he requires little else; and that if he uses them judiciously he will be able to combat, not without success, disease in its most unfavourable aspect. And although I would not have any one tie himself down to the precepts of this book alone, although it would be wrong rigidly to adhere to any one plan, for no one is perhaps always successful; although it is right for every one to judge, in some degree, for himself, and glean from every quarter what he thinks may be useful: still I am decidedly of the opinion that this is the book upon which the young tropical practitioner should place most reliance; that this will be his best guide at the outset. It is experience alone that will make the judicious and the confident practitioner: it is perhaps only the result both of fortunate and unfortunate cases that will inform us what is best; but until this expe rience has been gained, until we have our lessons from practice itself, we had as well, I think, lean fully more to the lessons of Johnson than any other. It is therefore a book which ought to be in the hands of all who go to India, or countries like it. And I venture to say, that he who studies it well, whatever his doubts and fears may be of his success in tropical practice, will find

M

himself armed, by what he draws from it, with no inconsiderable degree of confidence.

The last observation I have to make is, regarding the fitting out of the medicine chest. This may seem hardly necessary; but I do it because I have known it in some cases very badly managed. It sometimes happens that the surgeon is not appointed just until the ship is on the eve of sailing, and all the medicine is supplied before he joins her. Or, even though he has been appointed sometime before, he may think it just as well to allow the apothecary, whom the owners probably recommend, and who probably has supplied the ship before, to do the whole; just to let him supply what he is in the habit of supplying on like occasions. Then when he gets to sea, and comes to examine his medicine, he finds some articles that he probably never will use, and some he would wish much to have, he has not; others he may need a very little of, making a prominent bulk in the chest; and others he will need very much, dealt out rather sparingly. And thus when he is beyond the reach of remedying it, he finds himself very awkwardly circumstanced. It should therefore be a point with every surgeon to superintend the laying in of the medicine himself; and by doing so, he may not only save both himself and his patients considerable uneasiness, but he may also save the owners some considerable expense.

For a ship going a voyage averaging twelve months, and carrying fifty persons, I would lay in the following medicines. Magnes. sulph. ℔x. Sodæ sulph. ℔x; fol. sennæ, ℔ss.; sulph. sublim. ℔j.; super-tart. potas. ℔ij.; magnes. carbon. ℔ss.; pulv. rhei. ℥ij; pulv. jalap. ℥ij; ext. aloe, sp. ℥ij; ext. colocynth. c. ℥ij; pulv. g. gamboge ℥j; ol. ricini, ℔ij; hydrarg. submur. ℔ss.; pulv. antimon. ℥ij; pulv. ipecac. ℥ij; pulv. ipecac. c. ℥ij; antimon. tart. ℥ss.; opium, ℥ij; tinct. opii ℔j; æther sulph. ℥iv; æther nitros, ℥iv; liq. ammon. acet. ℔ss.; tinct. scillæ, ℥ij.; pulv. cinchon. ℔ij; rad. gentian. ℥iv; lig. quassiæ, ℔ss; flor. anthemid. ℔ss; sodæ subcarb. ℥iv; acid. tartar, ℥ij; acid.

sulphur. ℥iv; acid. nitric. ℔ss; camphor, ℥ij; ung. hydrarg. fort. ℔ij; ung. hydrarg. nit. ℔ss; ung. resin, ℔j; cerat. cetacei, ℔j; mass. pil. hydrarg. ℔j; bals. copaib. ℔ss; gum guaiac, ℥ij; zinc. sulph. ℥iv; cup.sulph. ℥ij; plumb. superacet. ℥iv; alum. ℥iv; nit. potass. ℥iv; cret. pp. ℥iv; tinct. catechu, ℥iv; liniment. sapo. c. ℔ss; ol. terebinth. ℔ss; ol. olivæ, ℔ss; liq. ammon. ℥iv; tinct. benzoin. c. ℥j; tinct. myrrh. ℥j; nit. argent. ℥ss; hydrarg. oxymur. ℥ss; hydrarg. oxyd. rub. ℥ss; liquor calcis, ℔j; emp. lyttæ, ℔ij; lint, ℔ss; adhesive strap, six yards; skin; court plaster. The usual appendages of the medicine chest.

When a considerable number of genteel passengers are to come under charge, who expect rather finer things occasionally than we can make the common sailor do with; and particularly if there be females and children amongst them, perhaps a few additional articles may be required, as, tinct. valerian. am. ℥iv; tinct. cardam. c. ℔ss; sp. levand. c. ℥ij; sp. menth. pip.

℥j; sp. anis. ℥j; pulv. calomb. ʒij; pulv. cinnam. ℥j; pulv. zinzib. ℥j; manna, ℔ss; confect. aromat. ʒij; tinct. assafœtid. ʒiv.

And these, I think, will in general be found quite enough both in number and quantity. Indeed some may suppose the collection rather a large one. But there is not one of the medicines, I will venture to say, but is very likely to be wanted sometime or other in the voyage; and what is the expense of two or three extra articles. What are a few pounds in comparison to the comfort which it may give both to the surgeon and his patient, when the scene of distress ensues, and there is no getting at the medicine market. Better, surely, that a little too much should be laid in; that a little even should be wasted, than that a life should be lost, or that an ailment should be allowed to continue which by some particular remedy might be alleviated or removed.

Some of the medicines may not keep well all the voyage; and of course, whenever we have reason to suppose that a medicine has lost, or is losing its virtue, we should set it aside and replenish. Drugs are in general plentiful enough in the Indian ports; so that any one which may spoil or run out, or any particular one wanted, can readily enough be procured, though at rather a high price.

THE END.

LONDON:
Printed by WILLIAM CLOWES,
Northumberland-court.

Zeitfracht Medien GmbH
Ferdinand-Jühlke-Straße 7
99095 Erfurt, Deutschland
produktsicherheit@kolibri360.de